Lectures on the Physics of Extreme States of Matter

Lectures on the Physics of Extreme States of Matter

Vladimir E Fortov

Joint Institute for High Temperatures, Moscow, Russia

IOP Publishing, Bristol, UK

ISBN   978-0-7503-2128-0 (ebook)
ISBN   978-0-7503-2126-6 (print)
ISBN   978-0-7503-2127-3 (mobi)

DOI   10.1088/2053-2563/ab1091

Version: 20190801

IOP Expanding Physics
ISSN 2053-2563 (online)
ISSN 2054-7315 (print)

British Library Cataloguing-in-Publication Data: A catalogue record for this book is available from the British Library.

Published by IOP Publishing, wholly owned by The Institute of Physics, London

IOP Publishing, Temple Circus, Temple Way, Bristol, BS1 6HG, UK

US Office: IOP Publishing, Inc., 190 North Independence Mall West, Suite 601, Philadelphia, PA 19106, USA

Contents

A word to the reader

You are holding a course of lectures from the 'Higher School of Physics' series of ROSATOM State Corporation.

The Higher School of Physics is the initiative of ROSATOM, aimed at training and educating scientists of a new generation in the field of theoretical and experimental physics, as well as at attracting talented young people to science and innovation spheres.

The books of this series have been prepared by leading scientists of the Russian Academy of Sciences and industry research centers and contain information about the most topical areas of theoretical and experimental physics.

I hope these books will become handbooks for students and postgraduates of specialized disciplines, young scientists and all employees of the nuclear industry, interested in improving their scientific and technical skills.

For ROSATOM, the matter of honor and professional maturity is to breathe fresh energy into nuclear power engineering and industry: to cultivate a galaxy of physicists of the future, who will become generators of innovative ideas and drivers of the world nuclear industry.

V A Pershukov,
Deputy Director General for Innovation Management,
ROSATOM State Corporation

From the editorial board

The successful history of the nuclear project, which was of key importance for the stability of our country for many decades was the result of the work of a huge team of scientists, engineers and workers. The task of forging an atomic shield was solved in a country destroyed by war, at the cost of incredible efforts, without a developed instrument-making infrastructure, in the absence of necessary unique materials and the corresponding industry. Paying tribute to all participants of the project, particular mention should be made of the decisive contribution of scientists. Bright representatives of physical, chemical, and materials sciences found solutions to the most complicated problems that were on the way to creating nuclear weapons. We proudly remember I V Kurchatov, Yu B Khariton, I E Tamm, A D Sakharov, K I Shchelkin, D A Frank-Kamenetsky, V L Ginzburg, E I Zababakhin and many other prominent scientists who led their colleagues and students. Success was determined by the talent and a broad knowledge of the leaders. Even today, their successors and disciples successfully work in our industry, including civil and defense spheres.

Modern problems of the development of science and technology also call for scientific leaders—custodians of the traditions initiated by previous generations. The upbringing of such leaders is the main concern of ROSATOM. That is why an idea arose to found the Higher School of Physics for young employees of ROSATOM institutes. The main task of the Higher School of Physics is to broaden the horizons of young people—students of the School by organizing four two-week modules on the basis of the largest scientific centers of ROSATOM, during which leading Russian scientists deliver lectures that represent different fields of physics and related sciences.

The selection of the courses and lecturers is made by the Scientific Council of the School. The Council includes well-known scientists from All-Russian Scientific Research Institute of Technical Physics (Snezhinsk), All-Russian Scientific Research Institute of Experimental Physics (Sarov), Troitsk Institute for Innovation and Fusion Research (Troitsk), and Institute of Physics and Power Engineering (Obninsk). Each course consists of six lectures; two courses are read each week; and the number of students is no more than 20 people, which creates prerequisites for direct contact of the lecturer with the audience.

It is important that students attend the courses only twice a year for two weeks. Young employees, who showed their qualities as researchers and leaders, are selected to join the School by the heads of the institutes.

The present series has been prepared on the basis of the lecture materials of the Higher School of Physics. The Scientific Council of the School expresses the hope that the series will appeal to a wide readership who wish to become acquainted with a brief summary of selected chapters of modern physics.

V P Smirnov,
Academician of the Russian Academy of Sciences,
Chairman of the Scientific Council of the Higher School of Physics of
ROSATOM State Corporation,
Chairman of the Editorial Board of the series

Author's preface

This book is based on the lectures delivered by the author at the Higher School of Physics founded by ROSATOM State Corporation, as well as on plenary, review and invited papers presented at scientific conferences and symposia.

I am grateful for the opportunity to acquaint students of the School with the current state and prospects for the development of the physics of extreme states of matter, with the advantages and limitations of various experimental methods of generation and diagnostics, and with the results achieved.

In this course, the author has made an attempt to systematize, summarize and present, from a single point of view, theoretical and experimental material relating to this new field of science. In addition to the extensive scientific literature, the author has used a large number of original papers, reports and abstracts that are not widely available to a wide audience.

In view of the vastness and dissimilarity of the material, the presentation is mainly of a fact-finding nature, referring the reader to relevant reviews and monographs. Therefore, many interesting astrophysical, laser and nuclear physics problems, as well as technical applications, are presented briefly, sometimes even schematically. The author, of course, did not set a goal to include everything that is known today about extreme states of matter. The emphasis is laid particularly on those issues that seem most interesting to the author and on which he and his colleagues had to work directly.

After the Introduction, the first lecture addresses the classification of states of matter at high energy densities. The general view of the phase diagram, dimensionless parameters and physical conditions corresponding to terrestrial and astrophysical objects are discussed.

The means for generating extreme states available to the experimenters are outlined in the second lecture.

The use of lasers to produce and diagnose states with high energy densities is considered in the third lecture.

The fourth lecture discusses the problems of the physics of extreme states of matter in collisions of heavy ions accelerated to sublight velocities, which are accompanied by the formation of superdense nuclear matter, i.e. compressed baryonic matter and quark–gluon plasma.

The problems relating to the description of the thermodynamics of a highly compressed electromagnetic plasma are addressed in the fifth lecture.

The book concludes with a discussion of the most characteristic astrophysical objects and phenomena associated with the implementation of extreme energy densities in the Universe under the action of gravity and thermonuclear energy release.

I hope that the book will be useful to a wide circle of scientists, postgraduates and students of natural-science specialties, providing access to original works and allowing them to unravel fascinating problems of modern physics of extreme states of matter.

The author will be grateful to readers for their critical comments, suggestions and amendments that are inevitable in presenting such a rapidly developing field as the physics of extreme states.

Vladimir E Fortov,
Academician of the Russian Academy of Sciences

Acknowledgements

In preparing the book, I had advice from and stimulating discussions with Yu Balega, N Andreev, S Blinnikov, M Kozintsev, A Sergeev, A Starostin, V Imshennik, I Iosilevskiy, N Kardashev, V Sultanov, B Sharkov, V Griaznov, and V Filinov, for which I am sincerely grateful.

Author biography

Vladimir E Fortov

Vladimir E Fortov is a member of the Russian Academy of Sciences, Head of the Department of Energy, Machinery, Mechanics and Control Processes of RAS.

Professor Fortov is an outstanding scientist who has made significant contribution to the physics of extreme states of matter and high energy densities, nonideal plasmas, shock and detonation waves, thermophysics, chemical physics, space research, and energetics, as well as to several other realms of physics and technology.

Vladimir E Fortov was born in 1946 in Russia, the town of Noginsk, Moscow Region. Fortov graduated from the Aerophysics and Space Research Department of the Moscow Institute of Physics and Technology (MIPT) with distinction in Thermodynamics and Aerodynamics and became a post-graduate student of MIPT. He defended his thesis, entitled 'Thermophysics of Nuclear Rocket Engines,' in 1971.

In 1971 Vladimir E Fortov started his research in the area of nonideal plasma physics and the thermophysical properties of extreme states of matter in the Chernogolovka Branch of the Institute of Chemical Physics of the USSR Academy of Sciences. The results of this research formed the basis for his doctoral thesis 'Nonideal plasma investigations using the dynamic method'.

In parallel with plasma research, Vladimir E Fortov was deeply involved in studies of the mechanics of deformation and damage to materials exposed to high pressures, temperatures, and high deformation rates.

The successful solution to many scientific problems was facilitated by Fortov's active cooperation with the General Physics Institute (GPI) and the Institute for High Temperatures (IHT) of the USSR Academy of Sciences. In 1994, a group supervised by Professor Fortov made a detailed prediction of the possible observable effects of an extraordinary space event—the collision of Shoemaker–Levy comet with Jupiter in July 1994. The data of subsequent observations carried out by many laboratories in the world confirmed the high accuracy of these predictions. Similar work was performed in 2005 in connection with the Deep Impact Project—a space experiment in which pioneering observations were made of a high-velocity collision of a metal striker with the nucleus of the 9P/Tempel comet.

In recognition of Vladimir E Fortov's work in the area of thermophysics and thermomechanics of extremely high pressures and temperatures, he was elected a corresponding member of the USSR Academy of Sciences in 1987 and a full member of the Russian Academy of Sciences (RAS) in 1991.

Another impressive line Professor Fortov's research is highly nonideal dust plasma. He supervised a series of pioneering experimental investigations into the structural and dynamic properties of plasma-dust crystals and liquids over a broad temperature and pressure range. For the first time, plasma crystals and liquids were obtained in a glow discharge, thermal plasmas, UV-radiation plasma, radioactive and cryogenic plasmas; experiments in plasma and crystallization under microgravity conditions were made aboard the Mir space station and the International Space Station (ISS).

Professor Fortov takes an active part in extreme expeditions. Specifically he participated in a cruise aboard the Volk atomic submarine; participated in the High-Latitude Arctic Deep-Sea Expedition to the North Pole; in the framework of the International Polar Year Program he took part in the International Antarctic Expedition to the South Pole and the Pole of Relative Inaccessibility; descended to the depths of Lake Baikal and Lake Leman (Switzerland) and visited the Vostok Polar Station in the Antarctic. Fortov rounded Cape Horn and the Cape of Good Hope on a yacht and crossed the Atlantic Ocean on a sailing yacht. He is keen on alpine skiing, tennis, piloting, and extreme traveling.

In recognition of his scientific and organizational activities, Vladimir E Fortov has been awarded many domestic and international prizes, a UNESCO Medal for his contributions to the development of nanoscience and nanotechnologies is among them. Numerous foreign and international academies and universities welcomed Vladimir E Fortov as their member.

Introduction

The states of matter at extremely high temperatures and densities have always attracted researchers, owing to the possibility of reaching record parameters, advancing to new domains of the phase diagram, and producing in the laboratory the exotic states that gave birth to our Universe through the Big Bang and which now account for the great bulk (90%–95%) of the mass of baryonic (visible) matter—in stellar and interstellar objects, planets, and exoplanets [1–9]. That is why the study of these states of matter—so exotic for us in terrestrial conditions and yet so typical for the rest of the Universe—is of great cognitive importance, forming our modern notions of the surrounding world.

Furthermore, a constant pragmatic incentive for such investigations is the application of highly compressed and heated matter in nuclear, thermonuclear, and pulsed power engineering, high-voltage and high-power electrophysics, for the synthesis of superhard materials, for strengthening and welding materials, for antimeteoritic protection of spacecraft and, of course, for defense. Indeed, the military application fostered the first successful experiment involving extreme states, which was conducted more than 3000 years ago—during the battle between David and Goliath. According to the Old Testament [10], the high-velocity impact of a stone shot from David's sling on Goliath's head killed him. It gave rise to a shock wave with an amplitude pressure of about 1.5 kbar. This pressure was more than twice the strength of Goliath's frontal bone and determined the outcome of the duel, to the great joy of the army and people of Israel. Discovered to be successful at that time, this scheme of action is today the ideological basis for all subsequent experiments in the field of dynamic physics of extreme states of matter.

Since the time of David, the application of more powerful and sophisticated energy cumulation systems—chemical and nuclear high explosives (HE), powder, light-gas, and electrodynamic guns, charged-particle fluxes, laser and x-ray radiation—has enabled the velocity of thrown projectiles to be raised by three to four orders of magnitude, and the pressure in the shock wave, by six to eight orders of magnitude, thereby reaching the megabar–gigabar pressure range and 'nuclear' energy densities in substances.

In the 20th century, the mainstream in physics of extreme states of matter was closely related to the entry of our civilization into the atomic and space era. In nuclear charges, extreme states of matter [11] generated by intense shock waves serve to initiate chain nuclear reactions in compressed nuclear fuel, and in thermonuclear charges and microtargets for controlled fusion, high-energy states are the main instrument for compressing and heating thermonuclear fuel and initiating thermonuclear reactions in it.

The research of extreme states of matter, starting in the mid-1950s within the framework of nuclear defense projects [12–16], has received considerable attention with the advent of new devices for generating high energy densities, such as lasers, charged particle beams, high-current Z-pinches, explosive-driven electric–discharge generators of high-power shock waves, multi-stage light-gas guns and diamond anvils. These sophisticated and expensive technical devices have made it possible to advance substantially along the scale of energy densities available for physical experiment and to obtain, in laboratory or quasi-laboratory conditions, the states of the megabar-gigabar pressure range unattainable for the traditional techniques of experimental physics.

Traditionally, energy densities are referred to as 'extreme' [1–4, 8] if they exceed 10^4–10^5 J cm^{-3}, which corresponds to the binding energy of condensed matter (for example, high explosives, hydrogen, or metals) and a pressure level of millions of atmospheres. For comparison, the pressures in the center of the Earth, Jupiter, and the Sun are about 3.6 Mbar, 40 Mbar, and 200 Gbar, respectively.

As a rule, matter in extreme states is in the plasma state—an ionized state arising from thermal- and/or pressure-induced ionization. In astrophysical objects, such compression and heating are caused by gravitational forces and nuclear reactions, and in laboratory conditions—by intense shock waves, which are excited by a wide variety of 'drivers', ranging from two-stage gas guns to lasers and high-current Z-pinches with a power of hundreds of terawatts[1]. However, while the lifetime of extreme states in astrophysical objects varies from milliseconds to billions of years, making it possible to conduct detailed observations and measurements with space probes and orbital and ground-based telescopes of different wavelengths, in terrestrial conditions we have to do with the microsecond–femtosecond–attosecond duration range [2, 3], which calls for the application of ultrafast specific diagnostic techniques.

At present, every large-scale physical facility (megaproject) that generates extremely high pressures and temperatures is enrolled in work programs (frequently international) on the fundamental physics of extreme states of matter, in addition to having practical, applied tasks in impulse energetics or defense. Thus, modern short-pulse laser systems (NIF, NIKE, USA; TRIDENT, LMJ, France; GEKKO-XII, Japan; OMEGA, VULKAN, Great Britain; Iskra-6, Russia; etc) are capable of releasing 1.0–1.8 MJ in a volume of the order of 1 mm^3 in several nanoseconds to produce pressures in the quasi-gigabar range (see tables 1.1 and 2.1).

In addition, the Z-pinch technology is now exhibiting considerable progress: at the Sandia facility (USA), $\approx$1.8 MJ soft x-ray radiation was obtained in the collapse of plasma liners during 5–15 ns in a region measuring about 1 cm^3. Supplemented by experiments with diamond anvils, explosion and electric explosion devices, and light-gas guns in the megabar pressure range, these record-high parameters are now the source of new and sometimes unexpected information about the behavior of highly compressed plasma [3].

Interestingly, in experiments on extreme-state laboratory plasma, even today it is possible to partly reproduce on a small scale many phenomena and processes occurring in astrophysical objects, information about which has become accessible due to the use of ground-based and spaceborne means of observation. These are the data on hydrodynamic mixing and various instabilities, shock-wave phenomena,

---

[1] The total power of Earth's electric power plants amounts to about 3.5 TW.

strongly emitting, relativistic and magnetized fluxes and jets, solitons, relativistic phenomena, equations of state, and the composition and spectra of compressed nonideal plasma, as well as the characteristics of interstellar cosmic plasma, dust, and a number of other effects.

Although the limiting pressures of a laboratory plasma are still 20–30 orders of magnitude higher than the maximum astrophysical values, this gap is being rapidly bridged, and the physical processes in a laboratory and in space often demonstrate an astonishing variety and at the same time striking similarities, evidencing at least the uniformity of physical principles of the behavior of matter in an extremely broad range of densities (approximately 42 orders of magnitude) and temperatures (up to 10^{13} K).

The revolutionary discoveries in astronomy of recent decades (neutron stars, pulsars, black holes, wormholes, γ-ray bursts, exoplanets, etc) [4–9] demonstrate new examples of extreme states, the investigation of which is important in order to solve the most fundamental problems of modern astrophysics.

To date, the physics of extreme states of matter has turned into an extensive and rapidly developing branch of modern science that makes use of the most sophisticated means of generation, diagnostic techniques, and numerical simulations using high-power supercomputers. It is no accident that half of the 30 problems of 'the physics minimum at the beginning of the 21st century' proposed by Academician V L Ginzburg [5] are to a greater or lesser degree dedicated to the physics of extreme states of matter.

The physics of extreme states of matter is closely related to several branches of science, including plasma physics and condensed-matter physics, relativistic physics, the physics of lasers and charged-particle beams, nuclear, atomic, and molecular physics, radiative, gas and magnetic hydrodynamics, astrophysics, etc. In this case, a distinguishing feature of the physics of extreme states of matter is an extreme complexity and strong nonlinearity of the physical processes occurring in it, the significance of collective interparticle interaction, and relativity, which makes the investigation of the phenomena in this field a fascinating and absorbing task, which attracts a constantly increasing number of researchers.

With all these reasons taken into account, the National Research Council of the US National Academies of Sciences formulated a large-scale national program of research [4] in the area of the physics of extreme states of matter and gave it high priority. Similar programs are being vigorously pursued in many developed countries capable of making the unique experimental devices and having qualified personnel in sufficient number.

The physics of extreme states of matter is a rapidly developing realm of modern science and technology, so that the material presented here will be permanently supplemented and improved by new measurements, observations, and models.

References

[1] Avrorin E N, Vodolaga B K, Simonenko V A and Fortov V E 1993 Intense shock waves and extreme states of matter *Phys.-Usp.* **36** 337–64

[2] Bible. Old Testament, 1, Samuel, 17: 34, 40, 43, 51.

[3] Drake R P 2006 *High-Energy-Density Physics* (Berlin: Springer)

[4] Fortov V, Iakubov I and Khrapak A 2006 *Physics of Strongly Coupled Plasma* (Oxford: Oxford University Press)

[5] Fortov V E 2007 Intense shock waves and extreme states of matter *Phys.-Usp.* **50** 333

[6] Henderson D (ed) 2003 *Frontiers in High Energy Density Physics* (Washington: National Research Council, Nat. Acad. Press)

[7] Ginzburg V L O 1995 *Fizike i Astrofizike (About Physics and Astrophysics)* (Moscow: Byuro Kvantum)

[8] Ginzburg V L 2004 On superconductivity and superfluidity (what I have and have not managed to do), as well as on the 'Physical Minimum' at the beginning of the 21st century (December 8, 2003) *Phys.-Usp.* **47** 1155

[9] Kirzhnits D A 1972 Extremal states of matter (ultrahigh pressures and temperatures) *Sov. Phys.-Usp.* **14** 512–23

[10] Rhodes R 1995 *Dark Sun; The Making of the Hydrogen Bomb* (New York: Simon and Schuster)

[11] Ryabev L D (ed) 1998 *Atomnyi Proekt SSSR: Dokumenty i Materialy (Atomic Project of the USSR: Documents and Materials). Vol. 1. 1938–1945. Part 1.* (Moscow: Fizmatlit)

[12] Ryabev L D (ed) 1999 *Atomnyi Proekt SSSR: Dokumenty i Materialy (Atomic Project of the USSR: Documents and Materials). Vol. 2. Atomnaya Bomba (Atomic Bomb). 1945–1954. Book 1* (Moscow: Fizmatlit)

[13] Ryabev L D (ed) 2000 *Atomnyi Proekt SSSR: Dokumenty i Materialy (Atomic Project of the USSR: Documents and Materials). Vol. 2. Atomnaya Bomba (Atomic Bomb). 1945–1954. Book 2* (Moscow: Fizmatlit)

[14] Vacca J R 2004 *The World's 20 Greatest Unsolved Problems* (Princeton, NJ: Prentice Hall)

[15] Vizgin V P (ed) 2002 *Istoriya Sovetskogo Atomnogo Proekta (History of the Soviet Atomic Project)* (St. Petersburg: Izd. RkhGU)

[16] Zasov A V and Postnov K A 2006 *Obshchaya Astrofizika (General Astrophysics)* (Fryazino: Vek-2)

IOP Publishing

Lectures on the Physics of Extreme States of Matter

Vladimir E Fortov

Chapter 1

Lecture 1: Matter under extreme conditions: classification of states

The scale of extreme states realized in Nature defies the most vivid imagination. At the bottom of the Mariana Trench (at a depth of 11 km), the water pressure, p, amounts to 1.2 kbar; in the center of the Earth, $p \approx 3.4$ Mbar, $T \approx 0.5$ eV, and the density, $\rho \approx 10$–20 g cm^{-3}; in the center of Jupiter, $p \approx 40$–70 Mbar, $\rho \approx 30$ g cm^{-3}, and $T \approx 2 \times 10^4$ K; in the center of the Sun, $p \approx 240$ Gbar, $T \approx 1.6 \times 10^3$ eV, and $\rho \approx 150$ g cm^{-3}; and in cooling-down stars (white dwarfs), $p \approx 10^{10}$–10^{16} Mbar, $\rho \approx 10^6$–10^9 g cm^{-3}, and $T \approx 10^3$ eV. In targets for controlled fusion with inertial confinement of plasma, $p \approx 200$ Gbar, $\rho \approx 150$–200 g cm^{-3}, $T \approx 10^8$ eV. Neutron stars, which are elements of pulsars, black holes, γ-ray bursts and magnetars, apparently have record-high parameters: $p \approx 10^{19}$ Mbar, $\rho \approx 10^{11}$ g cm^{-3}, and $T \approx 10^4$ eV for the mantle and $p \approx 10^{23}$ Mbar, $\rho \approx 10^{14}$ g cm^{-3}, $T \approx 10^4$ eV for the core at a giant induction of the magnetic field of 10^{11}–10^{16} Gs.

Collisions of heavy nuclei accelerated to relativistic velocities in modern accelerators lead to the emergence of supercompressed quark–gluon plasma states with ultra-extreme parameters $p \approx 10^{30}$ bar, $\rho \approx 10^{15}$–10^{16} g cm^{-3}, and $T \approx 10^{14}$ K, which exceed those realized in extreme astrophysical objects.

The emergence of extreme states in nature is due to the forces of gravity, which are inherently long-ranged and unscreened, unlike Coulomb forces (in electromagnetic plasma). These forces compress and heat the substance either directly or by stimulating exothermic nuclear reactions in massive astrophysical objects and in the early stages of the evolution of the Universe.

What is amazing is not only the breadth of the range of parameters realized in Nature, but also the huge difference in the characteristic times and dimensions. The dimensions of the visible part of the Universe amount to 1.3×10^{19} cm. The impression made by this figure becomes even stronger when it is compared with the time of 10^{-24} s taken by light to traverse a distance equal to about the

doi:10.1088/2053-2563/ab1091ch1

proton size (10^{-13} cm). The theory of relativity and other modern physical models do operate throughout this tremendous range.

As noted above, the lower boundary of the region of extreme states is considered to mean the states of a matter with an energy comparable to a binding energy of condensed matter, 10^4–10^5 J cm^{-2}, which corresponds to the binding energy of valence electrons (of several electron volts) and pressures from about 100 kbar to 1 Mbar. These pressures far exceed the ultimate mechanical strength of materials and make it necessary to take into account their compressibility during hydrodynamic motion under pulsed energy release.

In the domain of low pressures and temperatures, matter exhibits an exceptional diversity of properties and structures that we encounter daily under normal conditions [1].

Physical, chemical, structural, and biological properties of a substance under normal conditions are sharp nonmonotonic functions of the composition. The classification of these 'low-energy' states is complicated and cumbersome. It is determined by the position, details, and occupation features of electronic levels of atoms, ions, and molecules, and finally specifies the amazing richness of the forms and manifestations of organic and inorganic nature on Earth.

Laser and evaporative cooling methods (figure 1.1) enable ultralow (10^{-9} K) ion temperatures to be reached and interesting quantum phenomena such as Bose–Einstein condensation, Rydberg matter, Coulomb condensation, etc to be studied.

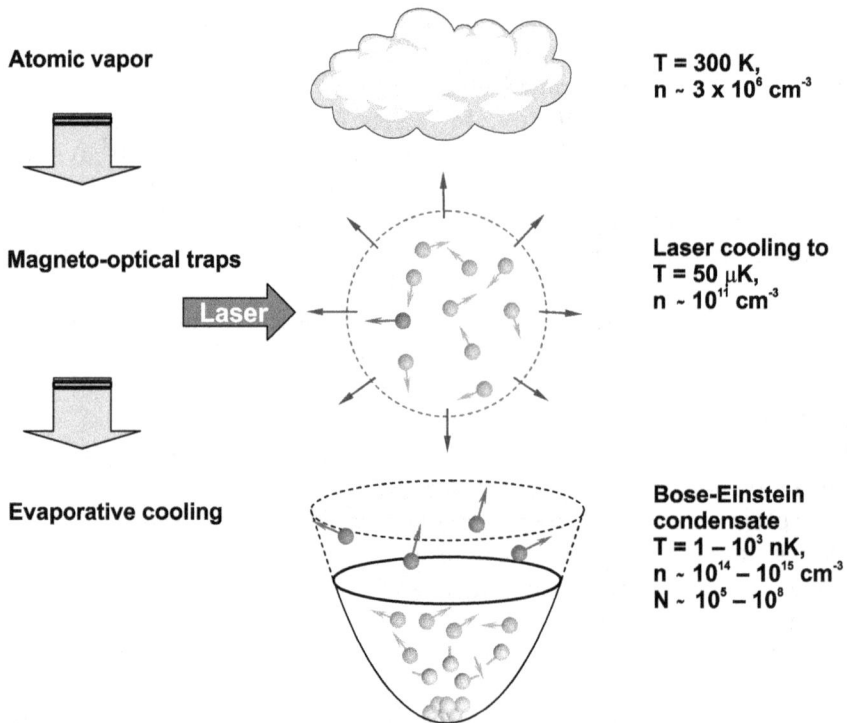

Atomic vapor

T = 300 K,
n ~ 3 × 10^8 cm^{-3}

Magneto-optical traps

Laser

Laser cooling to
T = 50 μK,
n ~ 10^{11} cm^{-3}

Evaporative cooling

Bose-Einstein condensate
T = 1 – 10^3 nK,
n ~ 10^{14} – 10^{15} cm^{-3}
N ~ 10^5 – 10^8

Figure 1.1. Methods for obtaining extremely low temperatures. Reprinted from [5] by permission from Springer. Copyright 2011, Springer.

With increasing energy density (p and T), substances acquire an increasingly universal structure [1–3]. The distinctions between the neighboring elements of the periodic system smooth out and the properties of a substance become progressively smoother functions of its composition. Owing to an increase in energy density, an obvious 'universalization' or simplification of the substance properties occurs. An increase in pressure and temperature ruptures molecular complexes to form atomic states, which then lose outer-shell electrons responsible for the chemical individuality of the substance, due to thermal and/or pressure-induced ionization. Electron shells of atoms and ions restructure to acquire an increasingly regular level occupation and a crystal lattice after a number of polymorphic transformations (this ordinarily takes place for $p < 0.5$ Mbar) transforms to a close-packed body-centered cubic structure common to all substances.

These processes of substance 'simplification' take place at energy densities comparable to the characteristic energies of the aforementioned 'universalization' processes. When the characteristic energy density becomes of the order of the valence shell energies, $e^2/a_0^4 \approx 3 \times 10^{14}$ erg cm^{-3} ($a_0 = \hbar/(me^2) = 5.2 \times 10^{-9}$ cm is the Bohr radius), the order of magnitude of the lower boundary of substance 'universalization', $T \approx 10$ eV, $p \approx 300$ Mbar, is reached. The exact quantitative determination of these boundaries is an important task of the experimental physics of extreme states of matter, especially due to the fact that theory [2, 3] predicts a highly varied behavior of substances in the ultramegabar pressure range (shell effects [2, 3], electron and plasma phase transitions [4–8] and other qualitative phenomena).

The upper boundary of the domain of extreme states is defined by the contemporary level of knowledge about the high-energy-density physics and observational astrophysical data, and is expected to be limited only by our imagination.

The ultra-extreme matter parameters available for modern physical concepts are defined by the so-called Planck quantities, which are combinations of the fundamental constants such as the Planck's constant $\hbar$, the velocity of light c, the gravitational constant G, and the Boltzmann constant k:

the length

$$l_p = \sqrt{\frac{\hbar G}{c^3}} = \frac{\hbar}{m_p c} \approx 1.62 \times 10^{-33} \text{ cm};$$

the mass (the so-called 'maximon' mass)

$$m_p = \sqrt{\frac{\hbar c}{G}} = 2.18 \times 10^{-5} \text{ g};$$

the time

$$t_p = \frac{l_p}{c} = \frac{\hbar}{m_p c^2} = \sqrt{\frac{\hbar G}{c^5}} = 5.39 \times 10^{-44} \text{ s};$$

the temperature

$$T_p = \frac{m_p c^2}{k} = \sqrt{\frac{\hbar c^5}{G k^2}} = 1.42 \times 10^{32} \text{ K};$$

the energy

$$W_p = m_p c^2 = \frac{\hbar}{t_p} = \sqrt{\frac{\hbar c^5}{G}} = 1.96 \times 10^9 \text{ J};$$

the density

$$\rho_p = \frac{m_p}{l_p^3} = \frac{\hbar t_p}{l_p^5} = \frac{c^5}{\hbar G^2} = 5.16 \times 10^{93} \text{ g cm}^{-3};$$

the force

$$F_p = \frac{W_p}{l_p} = \frac{\hbar}{l_p t_p} = \frac{c^4}{G} = 1.21 \times 10^{44} \text{ N};$$

the pressure

$$P_p = \frac{F_p}{l_p^2} = \frac{\hbar}{l_p^3 t_p} = \frac{c^7}{\hbar G^2} = 4.63 \times 10^{113} \text{ Pa};$$

the charge

$$q_p = \sqrt{\hbar c 4\pi\varepsilon_0} = 1.78 \times 10^{-18} \text{ C};$$

the power

$$P_p = \frac{W_p}{t_p} = \frac{\hbar}{t_p^2} = \frac{c^5}{G} = 3.63 \times 10^{52} \text{ W};$$

the circular frequency

$$\omega_p = \sqrt{\frac{c^5}{\hbar G}} = 1.85 \times 10^{43} \text{ s}^{-1};$$

the electric current

$$I_p = \frac{q_p}{t_p} = \sqrt{\frac{c^6 4\pi\varepsilon_0}{G}} = 3.48 \times 10^{25} \text{ A};$$

the voltage

$$U_p = \frac{W_p}{q_p} = \frac{\hbar}{t_p} = \sqrt{\frac{c^4}{G 4\pi\varepsilon_0}} = 1.05 \times 10^{27} \text{ V};$$

the impedance

$$Z_p = \frac{U_p}{I_p} = \frac{\hbar}{q_p^2} = \frac{1}{4\pi\varepsilon_0 c} = \frac{Z_0}{4\pi} = 29.98\ \Omega;$$

the electric field strength

$$E_p = \frac{U_p}{l_p} = \frac{1}{G}\sqrt{\frac{c^7}{4\pi\varepsilon_0\hbar}} = 6.4 \times 10^{59}\ \text{W cm}^{-1};$$

the magnetic field strength

$$H_p = \frac{1}{G}\sqrt{\frac{c^9 4\pi\varepsilon_0}{\hbar}} = 2.19 \times 10^{60}\ \text{A m}^{-1}$$
$$= 1.74 \times 10^{62}\ \text{Oe}$$

Such super-extreme parameters of matter, under which the known laws of physics seem to no longer work, might have been realized at the very beginning of the Big Bang or at the singularity in the collapse of black holes. In the first case, according to the model of the expanding Universe (A Friedman, G Lemaître [9, 10]), the Universe originated from the Planckian area of the order of 10^{-33} cm with ultra-high Planckian physical parameters and expanded to modern sizes of the order of 10^{28} cm over approximately 13.7–14.5 billion years. Here, owing to the gravitational compression of stars to the stage of black holes, singularities—ultrahigh parameters of the Planckian scale arise again. In these domains of singularities, physical models are now proposed according to which our space has more than three dimensions and that ordinary matter is in a three-dimensional manifold—the '3-brane world' [10]—embedded in this many-dimensional space. The capabilities of modern experiments in high-energy-density physics are far from these 'Planck' values and allow the properties of elementary particles to be elucidated up to energies of the order of 0.1–10 TeV and down to distances $\approx 10^{-16}$ cm.

Considering (following paper [1]) the energy range $mc^2 \approx 1$ GeV, which is amenable to a more substantial physical analysis and is nonrelativistic for nucleons, we obtain a boundary temperature of 10^9 eV, an energy density of 10^{37} erg cm^{-2}, and a pressure of about 10^{25} Mbar, although it is highly likely that even more extreme states of matter are realized in the cores of massive pulsars and could be found at early stages of the evolution of the Universe.

While our experimental capabilities are progressing rapidly, of course they are only partly able to encroach upon the region of ultra-extreme astrophysical states. Material strengths radically limit the use of static techniques for investigating high-energy densities, because the overwhelming majority of constructional materials are unable to withstand the pressures in question. The exception is the diamond—a record-holder in hardness ($\sigma_n \approx 500$ kbar); its use in diamond anvils allows a pressure of 3–5 Mbar to be reached in static experiments.

The palm of supremacy now belongs to dynamic techniques [7, 11, 12], which rely on the pulsed cumulation of high-energy densities in substances. The lifetime of such high-energy states is determined by the time of inertial plasma expansion, typically in the range 10^{-10}–10^{-6} s, which calls for the application of sophisticated fast diagnostic techniques. Physical conditions corresponding to the lower bound of states in question are listed in table 1.1 [9, 12, 13].

The phase diagram of the matter, corresponding to high-energy densities, is shown in figure 1.2 [9, 11, 12], which indicates the conditions existing in astrophysical objects as well as in technical and laboratory experimental devices. One can see that, being the most widespread state of matter in nature (95% of the mass of the Universe without dark matter), plasma occupies virtually the entire domain of the phase diagram. In this case, of special difficulty in the physical description of such a medium is the region of the nonideal plasma, where the Coulomb interparticle interaction energy $e^2 n^{1/3}$ is comparable to or exceeds the kinetic energy, E_k, of particle motion. In this domain, at $\Gamma = e^2 n^{1/3}/E_k > 1$, the effects of plasma nonideality cannot be described within the perturbation theory [1, 12], while the application of computer parameter-free Monte Carlo and molecular dynamics methods [4] is fraught with great difficulties of selection of adequate pseudopotentials and correct inclusion of quantum effects.

Table 1.1. Physical conditions corresponding to high energy densities of 10^4–10^5 J cm^{-3} [9].

Physical conditions	Values of physical parameters
Energy density W	$W \approx 10^4$–10^5 J cm^{-3}
Pressure p	$p \approx 0.1$–1.0 Mbar
Condensed high explosives:	$W \approx 10^4$ J cm^{-3}
pressure	≈ 400 kbar,
temperature	≈ 4000 K,
density	≈ 2.7 g cm^{-3},
detonation velocity	$\approx 9 \times 10^5$ cm s^{-1}
Impact of an aluminum plate on aluminum,	
velocity	$(5$–$13.2) \times 10^5$ cm s^{-1}
Impact of a molybdenum plate on molybdenum,	
velocity	$(3$–$7.5) \times 10^5$ cm s^{-1}
Electromagnetic radiation:	2.6×10^{15}–3×10^{15} W cm^{-2}
laser, intensity q $(W \sim q)$	2×10^2–4×10^2 eV
blackbody temperature T $(p \sim T^4)$	
Electric field strength E $(W \sim E^2)$	0.5×10^9–1.5×10^9 W cm^{-1}
Magnetic field induction B $(W \sim B^2)$	$1{,}6 \times 10^2$–5×10^2 T
Plasma density at temperature $T = 1$ keV	
$(p = nkT)$	6×10^{19}–6×10^{20} cm^{-3}
Laser radiation intensity q:	0.86×10^{12}–4×10^{12} W cm^{-2}
for $\lambda = 1$ μm, $W \sim q^{2/3}$	66–75 eV
blackbody temperature T $(p \sim T^{3,5})$	

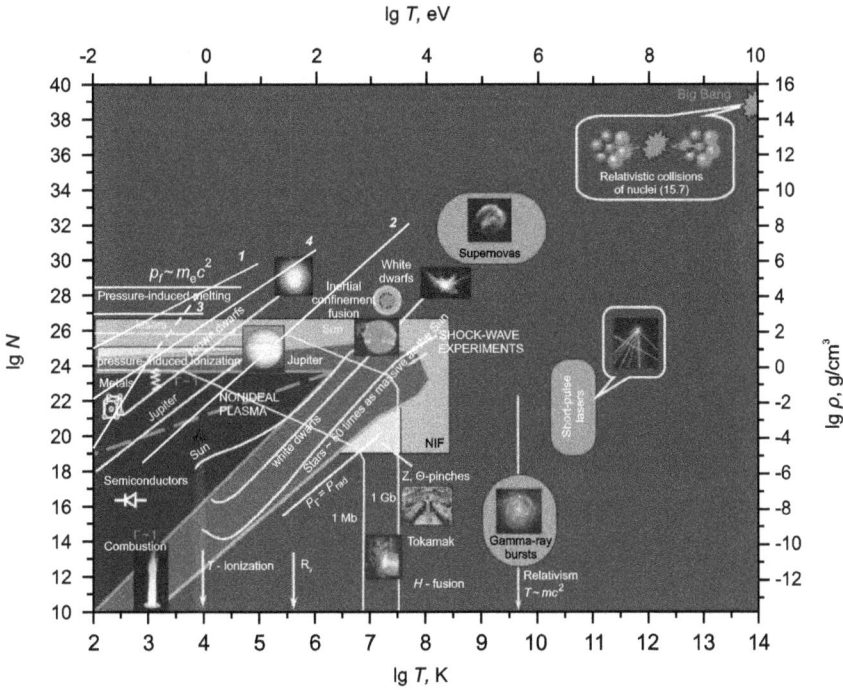

Figure 1.2. Phase diagram of states of matter [9, 11]. Curves *1–4* denote the states of nuclear matter component (neon) [1] on the lg ρ scale: *1*—boundary of the nucleus degeneracy region; *2*—boundary of the ideality region; *3*—melting curve; *4*—boundary of the region in which the lattice may be treated as classical. Reprinted from [5] by permission from Springer. Copyright 2011, Springer.

The effects of electron relativity in the equation of state and transport properties of the plasma, when $m_e c^2 \approx kT$, correspond to $T \approx 0.5$ MeV $\approx 6 \times 10^6$ K. Above this temperature, the matter becomes unstable with respect to spontaneous electron–positron pair production.

Quantum effects are determined by the degeneracy parameter $n\lambda^3$ ($\sqrt{\hbar^2/2mkT}$ is the thermal de Broglie wavelength). For a degenerate plasma, $n\lambda^3 \gg 1$, and the kinetic energy scale is the Fermi energy $E_F \approx \hbar^2 n^{2/3}/2m$, which increases with increasing plasma density, making it more ideal as it compresses, $n \to \infty$; $\Gamma = me^2/(\hbar^2 n^{1/3}) \to 0$. The relativity condition corresponding to $m_e c^2 \approx E_F \approx 0.5$ MeV yields a density $\rho \approx 10^6$ g cm^{-3}.

Similar asymptotics also takes place in another limiting case $T \to 0$ of a classical ($n\lambda^3 \ll 1$) plasma, where $E_k \approx kT$, and the plasma become more ideal [$\Gamma \approx e^2 n^{1/3}/(kT)$] upon heating. One can see that the periphery of the phase diagram is occupied by ideal ($\Gamma \ll 1$), Boltzmann ($n\lambda^3 \ll 1$), or degenerate ($n\lambda^3 \gg 1$) plasmas, which are described by the presently available adequate physical models [1, 4, 6, 11, 12].

The electron plasma in metals and semiconductors corresponds to the degenerate case with an interaction energy $E_{int} \sim e^2/r_0, |r_0 \sim \hbar|/k_F, E_k \sim k_F^2/m; \Gamma \sim e^2/\hbar v_F \approx 1-5$, where $v_F \sim 10^{-2}-10^{-3}$ s (s is the speed of light), and the subscript F refers to the parameters at the Fermi limit.

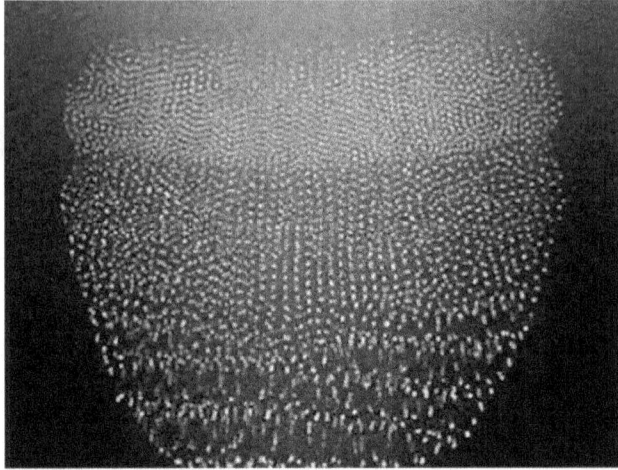

Figure 1.3. Plasma dust crystal and plasma liquid. Reprinted from [5] by permission from Springer. Copyright 2011, Springer.

For a quark–gluon plasma $E_{int} \sim g^2/r_0$, $r_0 \sim 1/T$, $E_k \sim T$; $\Gamma \approx 300$–400. For an ultracold plasma in traps, $\Gamma \sim (n/10^9)^{1/3}/T_k$. Most challenging for the theory is the vast domain of nonideal plasmas, $\Gamma \geqslant 1$, occupied by numerous technical applications (semiconductor and metal plasma, pulsed energetics, explosions, arcs, electric discharges, etc), where theory predicts qualitatively new physical effects (metallization, 'cool' ionization, dielectrization, plasma phase transitions, etc [11, 12]); the study of these effects requires substantial experimental and theoretical efforts.

Of special interest are plasma phase transitions in strongly nonideal Coulomb systems: crystallization of dust plasmas (figure 1.3) and ions in electrostatic traps and cyclotrons, in electrolytes and colloidal systems, and in two-dimensional electron systems on the surface of liquid helium, as well as exciton condensation in semiconductors, etc. Special mention should be made of the recently discovered phase transition in thermal deuterium plasma quasi-adiabatically compressed to megabar pressures by a series of reverberating shock waves.

The search for qualitatively new effects in the nonideal domain of parameters is a powerful and permanent incentive to investigate substances at high energy densities.

Another characteristic property of a high-energy-density plasma is the collective nature of its behavior and the strong nonlinearity of its response to external energy actions such as shock and electromagnetic waves, solitons, laser radiation, and fast particle fluxes. Thus, the propagation of electromagnetic waves in plasma excites several parametric instabilities (Raman, Thomson, and Brillouin scattering) and is accompanied by self-focusing and filamentation of radiation, by the development of inherently relativistic instabilities, by the generation of fast particles and jets, and—at higher intensities—by the 'boiling' of the vacuum with the electron–positron pair production.

Of special interest under extreme energy actions are transient hydrodynamic phenomena such as instabilities of shock waves and laminar flows, transition to the turbulent mode, turbulent mixing, and dynamics of jets and solitons.

Figure 1.4 borrowed from [9] shows the domains of the dimensionless parameters [Reynolds number, $Re \sim Ul/\nu$, and Mach number, $M = U/c$ (c is the velocity of sound, l is the characteristic size, and ν is the kinematic viscosity)], in which different hydrodynamic phenomena related to the physics of extreme states of matter are realized. The flow modes correspond to astrophysical applications, where $Re > 10^4$ and $M > 0.5$. In the explosion of a type Ia supernova, the Mach number ranges from 0.01 in the region of thermonuclear combustion to 100 in the shock wave arising due to the surface explosion.

All these fascinating and inherently nonlinear phenomena manifest themselves in both astrophysical and laboratory plasmas and, despite the enormous difference in spatial scale, have much in common and make up the subject of 'laboratory astrophysics'.

Laboratory astrophysics allows the states of matter and processes with high energy densities typical of astrophysical objects to be reproduced in microscopic volumes. These are the processes of instability and hydrodynamic mixing; ordinary and magneto-hydrodynamic turbulence; the dynamics of high-power shock, radiating, and soliton waves; expansion waves; magnetically compressed and fast relativistic jets; strongly radiating fluxes, and a number of other interesting and scarcely studied phenomena.

Of considerable interest is the information about the equation of state, composition, optical and transport properties, emission and absorption spectra, cross

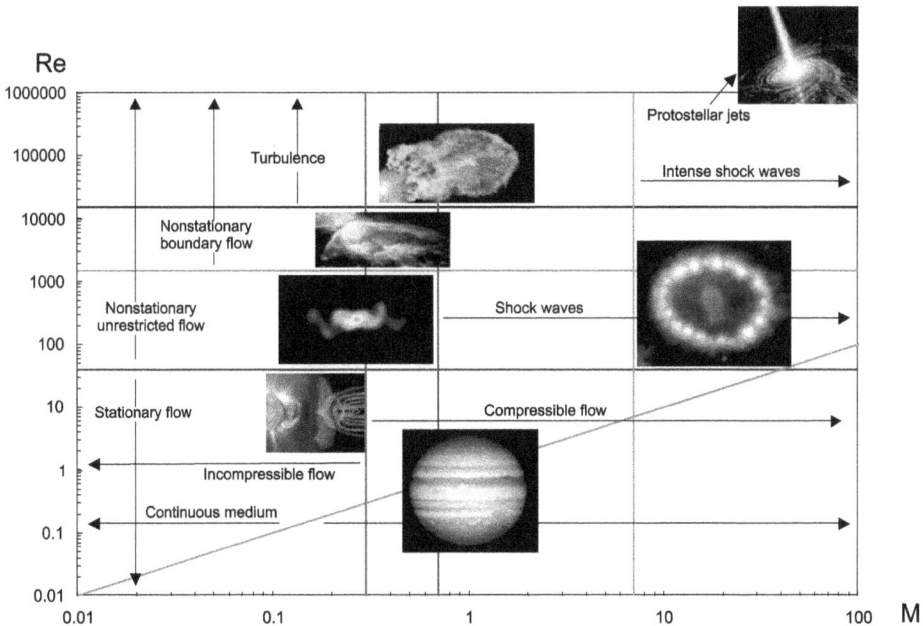

Figure 1.4. Hydrodynamic modes related to the physics of extreme states of matter. Reprinted from [5] by permission from Springer. Copyright 2011, Springer.

sections of elementary processes, radiation thermal conductivity coefficients, and properties of relativistic plasma. This makes it possible to study and model the physical conditions, stationary and pulsed processes in astrophysical objects and phenomena such as giant planets and exoplanets, stellar evolution and supernova explosions, gamma-ray burst structure, substance accretion dynamics in black holes, processes in binary and neutron stars as well as in the radiative motion of molecular interplanetary clouds, collisionless shock wave dynamics, charged-particle acceleration to ultrahigh energies, etc.

Let us now proceed to a more detailed description of the presently developed laboratory (lectures 2–4) and quasi-laboratory (section 2.2.4) methods for generating extreme states of matter.

References

[1] Caldirola P and Knoepfel H (ed) 1971 *Physics of High Energy Density* (New York: Academic)

[2] Fortov V, Iakubov I and Khrapak A 2006 *Physics of Strongly Coupled Plasma* (Oxford: Oxford University Press)

[3] Fortov V E (ed) 2000 *Entsiklopediya Nizkotemperaturnoi Plazmy (Encyclopedia of Low-Temperature Plasma)* (Moscow: Nauka)

[4] Fortov V E 2007 Intense shock waves and extreme states of matter *Phys.-Usp.* **50** 333

[5] Fortov V E 2011 Matter under extreme conditions: classification of states of matter *Extreme States of Matter* (Berlin: Springer)

[6] Fortov V E, Ivlev A V and Khrapak S A *et al* 2005 Complex (dusty) plasma: current status, open issues, perspectives *Phys. Rep.* **421** 1–103

[7] Fortov V E, Khrapak A G and Khrapak S A *et al* 2004 Dusty plasmas *Phys.-Usp.* **47** 447

[8] Henderson D (ed) 2003 *Frontiers in High Energy Density Physics* (Washington: National Research Council, Nat. Acad. Press)

[9] Kirzhnits D A 1972 Extremal states of matter (ultrahigh pressures and temperatures) *Sov. Phys.-Usp.* **14** 512–23

[10] Kirzhnits D A, Lozovik Y E and Shpatakovskaya G V 1975 Statistical model of matter *Sov. Phys.-Usp.* **18** 649–72

[11] Rubakov V A 2001 Large and infinite extra dimensions *Phys.-Usp.* **44** 871

[12] Shpatakovskaya G 2012 *Kvaziklassicheskii Metod v Zadachakh Kvantovoi Fiziki (Quasiclassical Method in Problems of Quantum Physics)* (Moscow: LAP LAMBERT Academic Publishing)

[13] Zel'dovich Y B and Raizer Y P 2002 *Physics of Shock Waves and High-Temperature Hydrodynamic Phenomena* (Mineola, NY: Dover)

IOP Publishing

Lectures on the Physics of Extreme States of Matter

Vladimir E Fortov

Chapter 2

Lecture 2: Extreme states of matter in laboratories

2.1 Main lines of research

The ultimate goal of experiments in macroscopic physics of extreme states of matter consists in the generation of states of matter, the parameters of which are at the boundaries of modern experimental capabilities. Even now, plasma states with peak pressures of hundreds to thousands of megabars, temperatures up to 10 billion degrees Celsius, and energy densities of 10^9 J cm^{-3}, which are comparable with the energy density of nuclear matter, have become the subject of laboratory investigations [1–3].

One of the pragmatic objectives of research in the physics of extreme states of matter is the controlled ignition of thermonuclear reactions. According to the ideas developed to date [4], to implement a controlled thermonuclear reaction with inertial hot plasma confinement requires an energy of several megajoules to be delivered to a spherical target in 10^{-9} s to generate at its center a deuterium–tritium plasma with extremely high parameters $T \approx (1\text{–}2) \times 10^8$ K, $\rho \approx 200$ g cm^{-3}, $p \approx 150\text{–}200$ Gbar, which is close to the conditions in the center of the Sun. Here, lasers are in the lead [5, 6], although electrodynamic techniques (Z-pinches) and heavy-ion schemes [7, 8] are being rapidly developed. The operation of such thermonuclear targets is basically close to supernovae explosions, allowing the vast wealth of experimental results and sophisticated computer codes created for the calculation of fusion microtargets and nuclear charges to be employed in astrophysics.

Apart from applications, of fundamental significance is the study of the equation of state of matter and plasma composition in a broad domain of the phase diagram, including the conditions inherent in giant planets, exoplanets, dwarfs, giants, and neutron stars. In thermodynamics, the determination of the quasi-classicality bounds (the Thomas–Fermi model [9, 10]) is still an important problem. Of considerable interest are the properties of degenerate compressed plasmas, their thermodynamics, equilibrium, kinetic, and transport properties under conditions of strong nonideality and in the presence of high-power magnetic fields, as well as the properties of a quark–gluon plasma.

Astrophysicists need reliable experimental data on the plasma properties at ultra-megabar pressures to construct and verify models of the structure and evolution of planets and exoplanets. For Jupiter and other planets, it is important to ascertain or disprove the existence of a hard core and determine the dimensions of the domain occupied by metallic hydrogen and the metallization bound for H_2 and $H_2 + He$. Of significance is the analysis of Jupiter's energetics with the allowance for phase layering of the mixtures He–H, C–O, etc, as well as the study of the origin and dynamics of its magnetic field. Similar problems are also encountered in studies of giant planets and exoplanets. In this case, quite important are shock-wave experiments, which enable the metallization bounds to be determined and the occurrence of a plasma phase transition to be ascertained.

The methods of laboratory energy cumulation now being developed will make it possible to obtain [11] relativistic jets and intense collisional, collisionless, and magnetohydrodynamic shock waves, much like those observed in astrophysical objects.

Projects are under way, involving the production of radiation-dominated hot plasmas, under conditions similar to those in black holes and accretion disks of neutron stars, as well as the stability of these modes (see the experiments involving nuclear explosions, lasers, and Z-pinches). Interesting suggestions are being put forward concerning the generation of radiatively collapsing magnetohydrodynamic and collisionless shock waves, fast particles, relativistic jets, and their focusing.

The modern experimental techniques for generating extreme states of matter open up interesting prospects for generating ultrahigh ($B > 1$ GG) magnetic fields and investigating their effect on the physical properties of matter. The fields obtained in laser-produced plasmas range already into hundreds of megagauss. Ultrahigh laser power levels may bring closer the prerequisites for the observation of relativistic gravitational effects.

2.2 Generators of high energy densities

The spectrum of experimental devices for generating high energy densities is highly diversified. It includes diamond anvils for static material compression, gunpowder and light-gas launching devices ('guns'), explosive-driven generators of intense shock waves, electroexplosion devices, magnetic cumulative generators, lasers, high-current generators of high-power electric current pulses, charged-particle accelerators, and possible combinations of these devices (see table 2.1).

Table 2.2 [11] compares the parameters of the highest-power facilities either in service today or under construction: lasers, pulsed electrical devices, Z-pinches, and charged-particle accelerators. Developed for plasma research in the interests of defense and high-energy physics, accelerators of relativistically charged particles are being successfully employed in basic plasma physics.

Table 2.1. Energy sources and experimental devices used in the physics of extreme states of matter [1].

Primarily energy source	Final form of the energy source	Energy density, MJ cm^{-3}	Temperature, eV	Pressure, 10^5 Pa	Total energy, MJ	Duration, s	Power, W
1	2	3	4	5	6	7	8
Chemical high explosives (HE)	Chemical HE	10^{-2}	0.5	5 × 10^5	10^2	10^{-7}	10^{10}
	Metal plates	0.3	60	10^7	3	10^{-6}	10^{10}
	1 MOe magnetic field	4 × 10^{-3}	0.3	5 × 10^4	5	10^{-6}	5 × 10^{12}
	25 MOe magnetic field	2.5	200	2.5 × 10^7	1	10^{-7}	10^{13}
	Explosion plasma generators	10^{-2}	60	10^5	30	10^{-6}	10^{12}
Nuclear HE	Nuclear HE	10^4	10^7	10^{10}	10^{11}	10^{-6}	10^{22}
	Neutron-induced heating	10	50	2 × 10^7	10^3	10^{-6}	10^{15}
Compressed gas	Shock waves in solids	5	50	5 × 10^7	10^4	3 × 10^{-6}	10^{15}
	Shock waves in gases	0.3	40	2 × 10^5	10^7	10^{-5}	10^{18}
	Adiabatic compression	2 × 10^{-5}	0.3	150	10^3	6 × 10^{-3}	10^5
	Pneumatic shock tubes	10^{-4}	1	250	10^{-2}	6 × 10^{-4}	3 × 10^8
	Combustion-driven shock tubes	10^{-6}	2	10	2 × 10^{-2}	3 × 10^{-4}	10^8
	Shock tubes, electric discharge	10^{-7}	2	1	10^{-2}	10^4	10^8
Capacitor	—	10^{-8}	—	—	40	10^{-5}	10^{12}
Rotor device	—	10^{-3}	—	—	100	10^{-4}	10^{12}
Inductive storage device	—	10^{-4}	—	—	100	10^{-4}	10^{12}
Storage battery	—	5 × 10^{-4}	—	—	1000	10^{-3}	10^{12}

(*Continued*)

Fast wire explosion	5×10^{-2}	4	10^5	10^{-3}	10^{-5}	10^9
Slow wire explosion	2×10^{-2}	0.5	5×10^2	10^{-3}	10^{-4}	10^7
Pulsed discharges	10^{-3}	10	10^4	10^{-4}	10^{-3}	10^9
Plasma focus	10^{-2}	10^3	10	10^{-4}	10^{-5}	10^{10}
High-pressure arcs	10^{-5}	2	10^4	10^{-4}	∞	10^4
Experiments in furnaces	10^{-3}	0.3	5×10^3	10^{-3}	∞	10^3
Laser —	10^{-6}	—	—	0.5×10^{-3}		10^{13}
Target	10^4	10^6	10^8	0.5	10^{-10}	5×10^{14}
Electron beam —	10^{-6}	—	—	1	10^{-8}	10^{14}
Target	5×10	5×10^3	10^7	0.1	10^{-8}	10^{13}

Table 2.2. Parameters of the facilities for generating high energy densities [11].

| | Laser facilities | | | Z-pinch | | | | | |
| | | | | Sandia | | C-300 | | Angara V | |
	NIF	LMJ	Petawatt laser	Current	X-rays	Current	X-rays	Current	X-rays
Energy per particle	3.6 eV	3.6 eV	1.5 eV	20 MA	50–250 eV	1.5–4 MA	70 eV	2–5 MA	100 eV
Pulse duration	1–20 ns	~10 ns	0.5 ps	100 ns (rise time)	5–15 ns	80 ns	12 ns	90 ns	6 ns
Spot size	0.3 mm	0.3 mm	5 μm	—	1 mm (cylinder)	—	2 mm	—	2 mm
Pulse energy	1.8 MJ	2 MJ	0.5–5 kJ	16 MJ	1.8 MJ	400 kJ	50 kJ	600 kJ	120 kJ
Intensity (W cm^{-2})	2×10^{15}	~10^{15}	10^{22}	—	10^{14}	—	2–3×10^{12}	—	6–10×10^{12}

Accelerators

| | Electron accelerator (SLAC) | LHC accelerator | SIS 18 | | SIS 100 | TWAC |
			Today	Design		
Energy per particle	50 GeV	7 TeV	1 GeV	1 GeV	4 GeV	700 MeV
Pulse duration	5 ns	0.25 ns	200 ns	50 ns	20 ns	100 ns
Spot size	3 mm	16 mm	1 mm	1 mm	1 mm	1 mm
Pulse energy	150 J	334 MJ	~1 kJ	30 kJ	300 kJ	100 kJ
Intensity (W cm^{-2})	10^{20}	10^{19}	5×10^{11}	6×10^{13}	10^{15}	10^{14}

2.2.1 Static methods using diamond anvils

Significant progress in the area of static pressures was achieved in the early 1980s with the advent of the experimental technique of diamond anvils (figure 2.1). In these facilities, two specially faceted diamonds compress thin (10–100 μm) plane layers of the substance under investigation to the highest attainable pressures of the megabar range, their upper limit being defined by the ultimate strength of diamond, equal to about 0.5–1.0 Mbar. In a number of experiments, the compressed substance is heated by laser radiation (figure 2.2), or the material compressed in diamond anvils is the target for its subsequent compression by laser-driven shock waves.

The unlimited time of static compression permits a wide spectrum of diagnostic tools to be employed, including different kinds of spectroscopy, and also x-ray structure analysis to be performed with the use of kiloelectronvolt x-ray and synchrotron radiation. Experiments of this kind have yielded a wealth of useful information about the mechanical properties, thermodynamics, and phase transformations in geophysical objects (figure 2.3) in the parameter range $p \approx 0.1$–3.6 Mbar, $T \approx 10^3$–6×10^3 K, which is extreme for terrestrial conditions.

The capabilities of diamond anvil techniques are limited by the strength of diamond (the strongest terrestrial material) and will hardly go beyond the megabar pressure range.

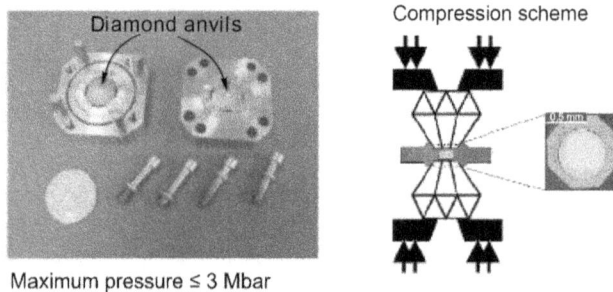

Figure 2.1. Facility for static compression of a substance in diamond anvils. Reprinted from [18] by permission from Springer. Copyright 2011.

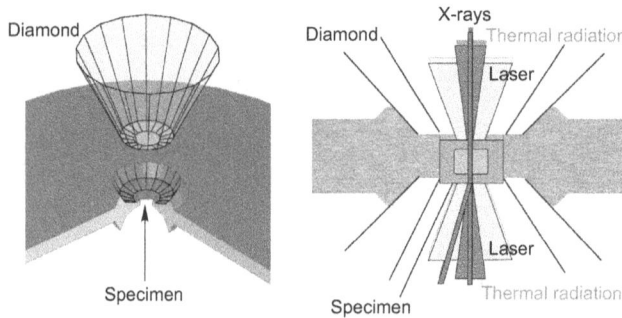

Figure 2.2. Schematic representation of a static experiment on compression of a substance in diamond anvils involving laser heating.

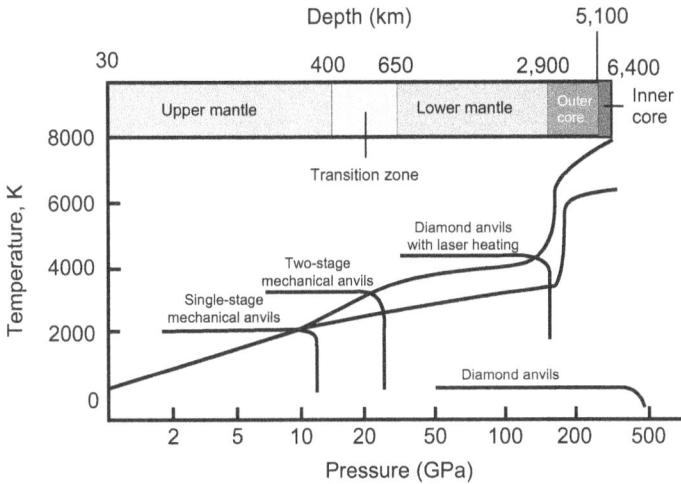

Figure 2.3. Comparison of the parameters attainable under static conditions and the physical conditions in the Earth's interior. Reprinted from [18] by permission from Springer. Copyright 2011.

2.2.2 Dynamic methods

Further advancement towards high energy densities is related to the transition to dynamic methods of investigation [1–3, 12–14], which are based on pulsed energy cumulation in a substance in question by means of intense shock waves or by means of electromagnetic or corpuscular radiations of different types. The plasma temperatures and pressures arising in this case dramatically exceed the thermal and mechanical strengths of the structural materials of the facilities, resulting in limitations on the characteristic plasma lifetime in dynamic experiments, which is determined by the target expansion dynamics and is equal to $\approx 10^{-10}$–10^{-5} s. In the dynamic approach there are no fundamental limitations on the maximum energy densities and pressures produced in the target: they are limited only by the power of the energy source, i.e. the 'driver'.

The routine tool for producing high energy densities is high-power shock waves [12, 13], which emerge due to nonlinear hydrodynamic effects in a substance during its motion caused by a pulsed energy release. In this case, a major part is played by the shock wave, a viscous compression shock, in which the kinetic energy of the oncoming flux is converted into thermal energy of the compressed and irreversibly heated plasma.

The shock-wave technique plays a leading role in high-energy-density physics today, making it possible to produce maximum pressures of the megabar and gigabar ranges for many chemical elements and compounds. The current range of peak dynamic pressures is six orders of magnitude higher than those occurring upon the impact of a bullet and three orders of magnitude higher than those in the center of the Earth, and is close [2, 3] to the pressure in the central layers of the Sun and inertial thermonuclear fusion targets. These exotic states of a substance emerged during the birth of our Universe, within several seconds after the Big Bang [11, 15, 16].

Shock waves not only compress a substance, but also heat it to high temperatures, which is of particular importance for the production of a plasma, i.e. the ionized state of matter. A number of dynamic techniques is being employed to experimentally study strongly nonideal plasmas [1, 12, 14] (figure 2.4).

The shock compression of an initially solid or liquid substance enables states of nonideal degenerate (Fermi statistics) and classical (Boltzmann statistics) plasmas compressed to peak pressures of ≈ 4 Gbar and heated to temperatures of $\approx 10^7$ K [2, 3] to be produced behind the shock front; at these parameters, the density of the inertial plasma energy is comparable with the nuclear energy density, and the temperatures are close to the conditions under which the energy and pressure of equilibrium radiation play a noticeable role in the total thermodynamics and dynamics of such high-energy states.

To reduce the irreversible heating effects, it is expedient to compress a material by a sequence of incident and reflected H_k shock waves, when compression becomes closer to the 'softer' isentropic compression, making it possible to obtain substantially higher compression ratios (10–50 times) and lower temperatures (≈ 10 times) in comparison with a single-stage shock-wave compression.

Multiple shock compression has been used successfully for the experimental study of pressure-induced plasma ionization and substance dielectrization [17] at megabar pressures. Quasi-adiabatic compression S_1 has also been realized in the highly symmetric cylindrical explosive compression of hydrogen and inert gases (figure 2.5). The highest plasma parameters were obtained using spherical explosive compression [18].

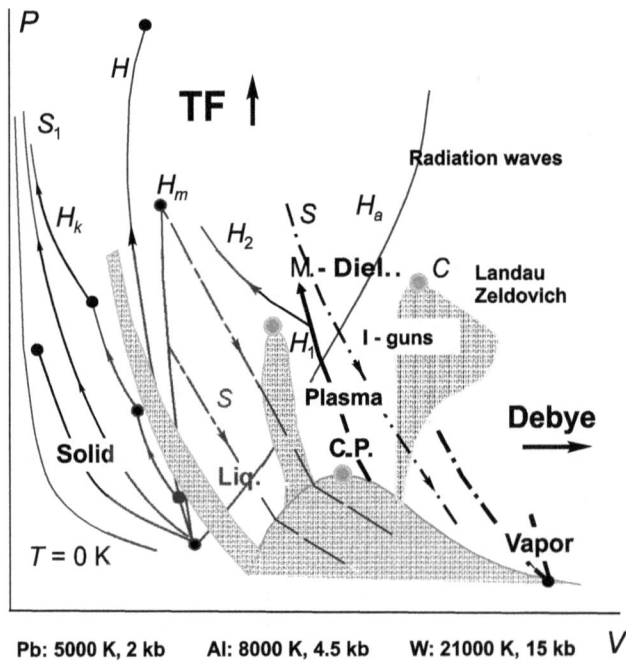

Pb: 5000 K, 2 kb   Al: 8000 K, 4.5 kb   W: 21000 K, 15 kb   V

Figure 2.4. Thermodynamic trajectories of dynamic substance investigation techniques [1]. The critical point (CP) parameters of several metals are given at the bottom.

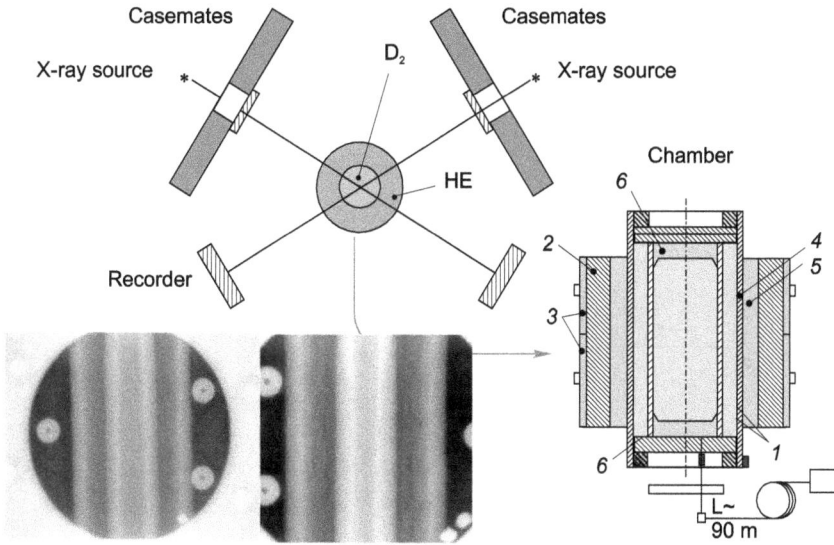

Figure 2.5. Cylindrical explosion devices for quasi-adiabatic plasma compression: 1—cylindrical specimen; 2—HE charge; 3, 4—external and internal metal liners; 5—x-ray radiation source; 6—x-ray recorders. Reprinted from [18] by permission from Springer. Copyright 2011.

The experiment was performed using an x-ray complex of three betatrons and a multichannel optoelectronic system for recording the x-ray images of the process of spherical deuterium compression.

In another limiting case, when a high-temperature plasma is required, it is expedient to subject lower (in comparison with solid) density targets, for example, porous metals H_m or aerogels H_a (see figure 2.4) to shock-wave compression. This makes it possible to sharply strengthen the irreversibility effects of shock compression and thereby increase the entropy and temperature of the compressed state.

Figure 2.6 shows experimental data on the thermodynamics of high-energy states in the (unconventional for plasma physics) range of solid-state densities and high temperatures obtained by shock-wave compression of porous nickel samples. Interestingly, these experimental data correspond to the metal–dielectric transition region (figure 2.7), where pressure-induced and temperature ionization effects are significant for the description of plasma thermodynamics [1, 19].

The shock compression of noble gases and saturated alkali metal vapors by incident H_1 and reflected H_2 shock waves (see figure 2.4) allows the plasma to be studied in the domain with developed thermal ionization, where the electrons obey Boltzmann statistics [1, 14].

A characteristic feature of the shock-wave technique is that it permits high pressures and temperatures to be obtained in compressed media, while the low-density domain (including the boiling curve and the vicinity of the critical point) turns out to be inaccessible to it. The plasma states intermediate between a solid and a gas are studied using the isentropic expansion technique based on the generation of a plasma in the adiabatic expansion S of a condensed substance precompressed and irreversibly preheated at the front of an intense shock wave [1, 14].

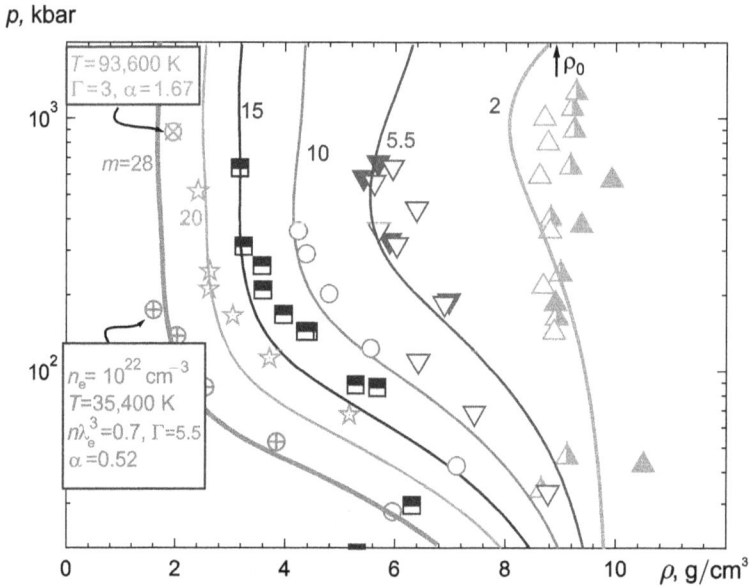

Figure 2.6. Thermodynamics of a nonideal nickel plasma. Symbols are the results of shock compression of porous ($m = \rho_0/\rho_{00}$) specimens; α is the degree of ionization. Reprinted from [18] by permission from Springer. Copyright 2011.

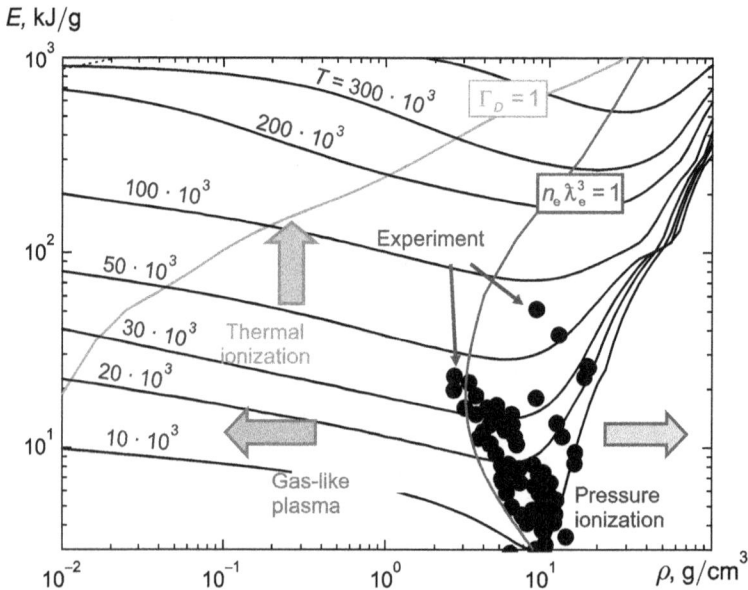

Figure 2.7. Energy density of a shock-compressed nickel plasma. Reprinted from [18] by permission from Springer. Copyright 2011.

This technique was first used to experimentally study the high-temperature portions of the boiling curves, the near-critical states, and the metal–dielectric transition domains for a large number of metals. As an example, figure 2.8 shows the

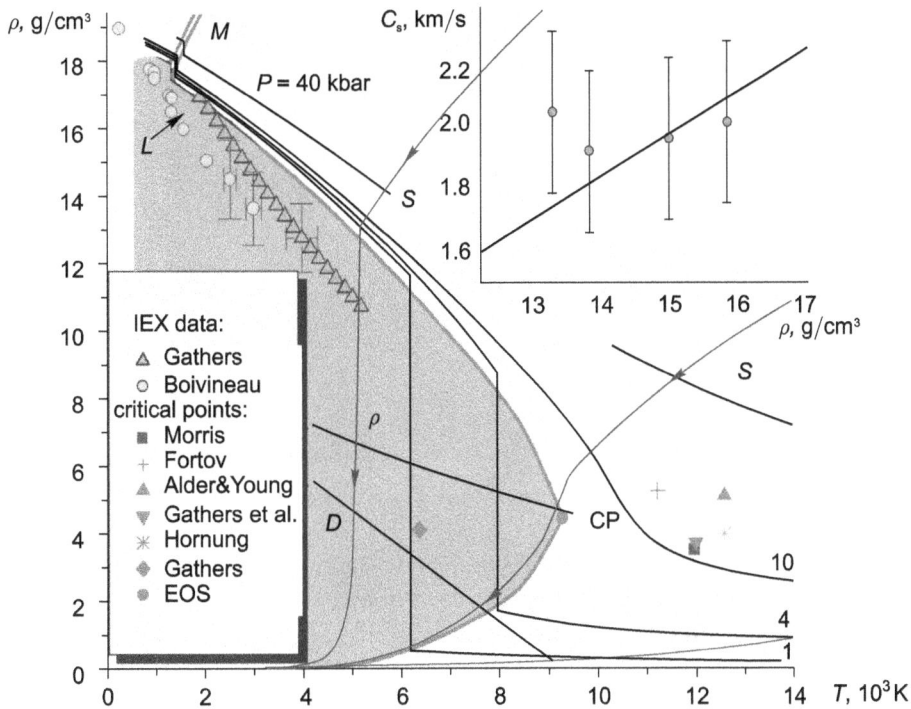

Figure 2.8. High-temperature evaporation of uranium in the near-critical domain. The data was obtained using the adiabatic expansion method. Reprinted from [18] by permission from Springer. Copyright 2011.

domain of high-temperature evaporation of uranium, which was obtained by using this data [1, 14].

Thus, dynamic techniques in their different combinations permit a broad spectrum of plasma states with a variety of strong interparticle interactions to be realized experimentally and investigated. In this case, it is possible not only to experimentally realize conditions with high energy densities, but also to diagnose sufficiently completely these high-energy states, since shock and adiabatic waves are not only a means of generation, but also a specific tool for diagnosing extreme states of matter with a high energy density [1, 12–14]. Measurements of the mechanical parameters of the motion of shock waves and contact discontinuities make it possible to determine the thermodynamic plasma properties and, with the use of modern high-speed diagnostic techniques, many physical characteristics of extreme parameter plasmas.

Let us now briefly discuss the experimental technique for generation of shock waves in dense media.

2.2.3 Light-gas guns and chemical high explosives

Today the technique of high-power shock waves generated by the impact of metal liners (strikers) accelerated to velocities of several kilometers per second on a target of the substance in question is the main source of physical information about the

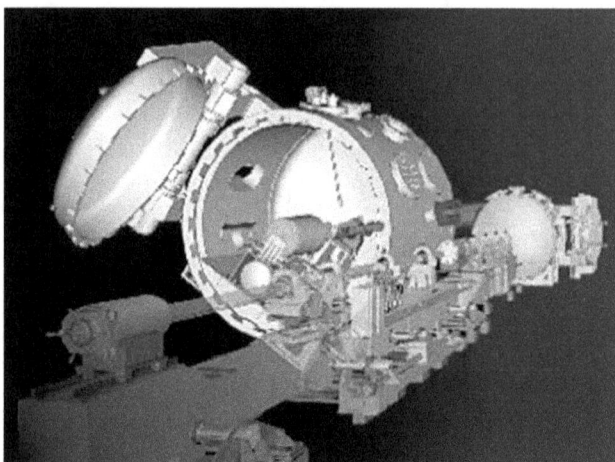

Figure 2.9. Schematic representation of a light-gas gun, Livermore, USA.

plasma behavior at pressures up to 10–15 Mbar. Here we shall not describe in detail the liner acceleration technique and the means of diagnostics—they are dealt with in comprehensive reviews and monographs [1, 2, 12, 14, 20, 21]. Note only that in shock-wave experiments of this kind it is possible to carry out sufficiently ample measurements of the physical plasma properties. The equation of state is determined by electrocontact and optical recording of the time intervals in the motion of shock-wave discontinuities and contact surfaces. Pyrometric, spectroscopic, protono-graphic, x-ray diffraction, and adsorption measurements are performed using pulsed x-ray and synchrotron radiation sources; laser interferometric measurements are made; low- and high-frequency Hall conductivities are recorded; and piezo- and magnetoelectric effects are detected.

Gunpowder and light-gas launching devices—'guns' (figure 2.9)—were most widely used in the USA, while in the USSR preference was given to explosive propellant devices [20].

To increase the velocity of launching and hence the shock-compressed plasma pressure, use is made of highly sophisticated gas-dynamic techniques. Thus, the method of 'gradient' cumulation (figure 2.10 [22]) is based on a successive increase in the velocity of strikers in alternating heavy and light material plane layers. This method is not related to the effects of geometrical energy focusing and therefore exhibits a higher stability of acceleration and compression in comparison with the spherical one. The thus-obtained three-stage 'layer cake' explosion [20] accelerates a 100 μm molybdenum striker to velocities of 5–14 km s^{-1}, thereby exciting in a target plasma a plane shock wave or a series of reverberating shock waves with megabar amplitude pressures. The geometrical parameters of these experimental devices are selected in such a way as to eliminate the distorting effect of side and rear unloading waves and to ensure the one-dimensionality and stationarity of gas dynamic flow in the region of recording.

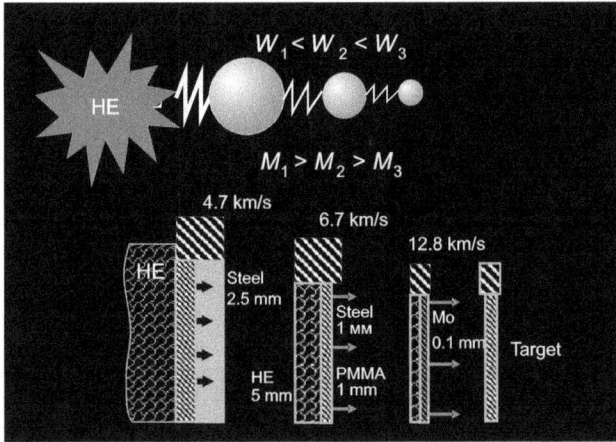

Figure 2.10. Principle of 'gradient' cumulation [22] and three-stage 'layer cake' explosion [12].

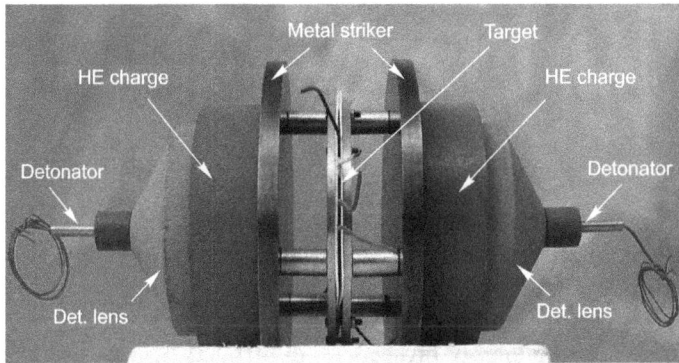

Figure 2.11. Explosive-driven generator of counterpropagating shock waves.

To increase the parameters of shock compression, several experiments made use of explosive-driven generators of counterpropagating shock waves (figure 2.11), where the material under study was loaded on both sides by a synchronous impact of steel strikers symmetrically accelerated by high explosive (HE) charges.

High-precision spherical explosive-driven generators of intense shock waves (figure 2.12) were made in the USSR [12, 20] to study thermodynamic material properties at pressures ranging up to 10 Mbar. Using the geometrical cumulation effects in the centripetal motion (implosion) of detonation products and hemispherical shells, it was possible to accelerate metal strikers to velocities of the order of 23 km s^{-1} in devices weighing about 100 kg with an energy release of about 300 MJ.

The experimental technique of high-power shock waves for studying extreme states of matter [11, 23] is today the main source of information about the behavior of strongly nonideal heavily compressed plasmas in the domain of record-high

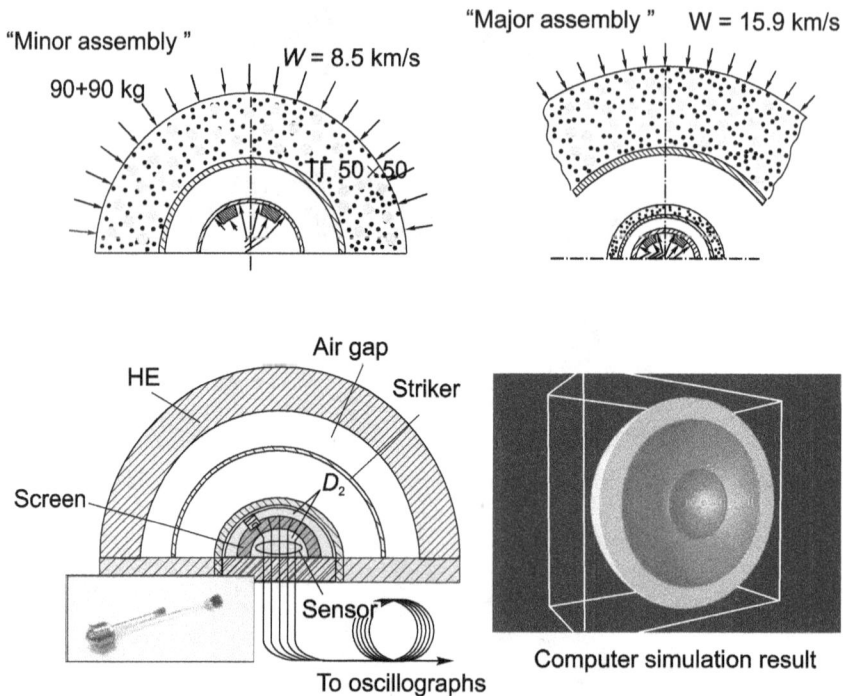

Figure 2.12. Explosive-driven generators of spherically converging intense shock waves.

temperatures and megabar-to-gigabar pressures. Being exotic for terrestrial conditions, these ultra-extreme states are quite typical of the majority of astrophysical objects and define the structure, evolution, and luminosity of stars, the planets of the solar system and recently discovered exoplanets.

In addition, promising energy projects on controlled fusion with inertial plasma confinement and the realization of high-temperature states in compressed hydrogen are associated with plasmas of ultramegabar range. These circumstances are a permanent factor that gives an impetus to experimental studies into the properties of heavily compressed nonideal plasmas of hydrogen, deuterium, and inert gases by high-power shock waves, which are excited by light-gas and explosion plane or hemispherical devices, high-power lasers, and electrodynamic accelerators [11, 23].

In higher-stability conical explosive-driven generators, use is made of cumulation effects in the irregular ('Mach') convergence of cylindrical shock waves (SW) (figure 2.13). The combination of irregular cylindrical and 'gradient' cumulation effects enables a shock wave to be excited in copper with an amplitude of ≈ 20 Mbar, which is comparable with pressures in the near zone of a nuclear explosion.

2.2.4 Underground nuclear explosions

Plasma parameters that are record-high for terrestrial conditions were obtained in the near zone of a nuclear explosion. Schemes of several experiments are presented in figures 2.14 and 2.15 [2, 3, 20, 21]. The combination of experimental data on a

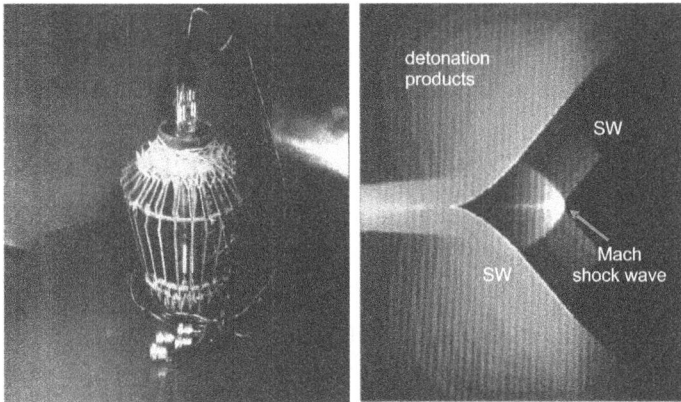

Figure 2.13. Explosive-driven generators of conically converging 'Mach' shock waves. Shown on the right are the results of a two-dimensional hydrodynamic simulation.

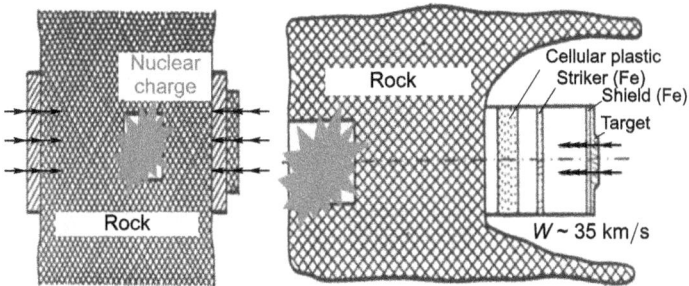

Figure 2.14. Schematic representation of experiments on generation of high-power shock waves in the near zone of a nuclear explosion. Reprinted from [18] by permission from Springer. Copyright 2011.

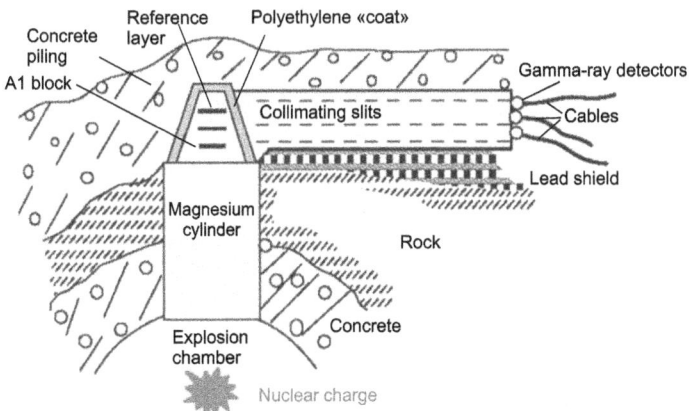

Figure 2.15. Schematic representation of experiments with an underground nuclear explosion using a gamma-active reference layer [3]. Reprinted from [18] by permission from Springer. Copyright 2011.

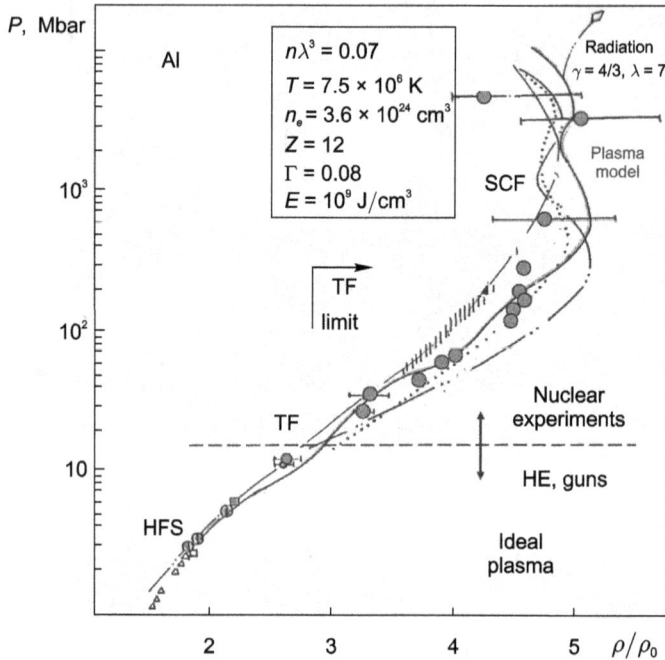

Figure 2.16. Shock-wave compression of aluminum to gigabar pressures. Reprinted from [18] by permission from Springer. Copyright 2011.

shock-compressed aluminum plasma is shown in figure 2.16, where the highest points correspond to record-high parameters in terrestrial conditions.

The density of the internal energy of this plasma is $E \approx 10^9$ J cm^{-3}, which is close to the energy density of nuclear matter, and the pressure $p \approx 4$ Gbar is close to the pressure in internal layers of the Sun. The plasma under these conditions ($n_e \approx 4 \times 10^{24}$ cm^3, $T \approx 8 \times 10^6$ K) is nondegenerate, $n\lambda^3 \approx 0.07$; moreover, it is twelve times ionized, and the nonideality parameter is small ($\Gamma \sim 0.1$), which is an experimental illustration of the thesis (lecture 1) about simplification of the physical plasma properties in the limit of ultrahigh energy densities. Interestingly, the parameter range in question is adjacent to the domain where the energy and pressure of equilibrium light radiation make an appreciable contribution to the thermodynamics of the system:

$$E_R = 4\sigma T^4/c; \quad p_R = E_R/3 = 4\sigma T^4/(3c).$$

Thus, the plasma dynamics mode is realized, which is close to the radiative gas-dynamic one.

The pressures realized by means of nuclear explosions [2, 3, 20, 21] (figure 2.16) belong to the multimegabar pressure domain and are close to the characteristic 'physical' pressure, which may be found from dimensionality considerations $p \approx e^2/a_B^4 \approx 300$ Mbar ($a_B = \hbar^2/(me^2)$ is the Bohr electron radius); starting with these pressures, the Thomas–Fermi model [9, 10] becomes applicable, which implies a simplified quantum-statistical description of a strongly compressed substance and the 'self-similarity' of its physical properties. This model [10], which relies on the quasi-classical approximation to the self-consistent field (SCF) method, is a substantial simplification of the many-particle quantum-mechanical problem and therefore enjoys wide use in the solution of astrophysical and special problems.

The physical conditions for the applicability of the quasi-classical model, as we have noted, correspond to extremely high pressures $p \gg 300$ Mbar and temperatures $T \gg 10^5$ K, which are realized in various astrophysical objects, but so far are not easily accessible for experimental techniques under terrestrial conditions. The currently attainable states in the region of high pressures and temperatures are realized now, as we have seen above, with the help of dynamic methods using the technique of intense shock waves. Although the majority of shock-wave experimental data do not correspond exactly to quantum-statistical conditions, they permit the properties of the quasi-classical models to be extrapolated beyond the limits of their formal applicability defined [10, 24] by the smallness of the corresponding literal criteria. The results of these constructions show that the introduction of quantum, exchange, and correlation corrections (oscillation corrections were not included) improves extrapolation, which becomes possible in this case to pressures $p \geqslant 300$ Mbar for zero temperature and to about 50 Mbar at $T \geqslant 10^4$ K. At the same time, the ambiguous interpretation of the results of comparative measurements performed in underground nuclear explosions does not give a definite answer to the question of the preferability of one or another variant of the quasi-classical model; the results of these measurements contradict the absolute measurement data [3].

The discrepancy between the calculations by the Thomas–Fermi model and the more accurate quantum mechanical calculations by the Hartree–Fock–Slater (HFS) model in the characteristic domain for the plasma state was considered in paper [1].

It should be noted that at present the question of the limits of applicability of the quasi-classical model remains largely open, and the nature of the behavior of the substance at $p > 300$ Mbar turns out to be more diverse than was previously assumed on the basis of simplified representations [10]. Experimental verification of the predictions of the quasi-classical shell model is presently the most interesting problem of ultrahigh-pressure physics; solving this problem will obviously call for new experimental techniques basing on high-power directional energy laser fluxes.

For the time being, only the underground nuclear explosion techniques [2, 3, 20] furnish the opportunity to approach multimegabar pressures, making it possible to estimate the lower bounds of validity of the quasi-classical model. According to [12], this model is applicable, starting with pressures of about 100 Mbar on the Hugoniot curve, whereas the range of its applicability is noticeably narrowed with increasing temperature (shock adiabats of a porous substance).

The thermodynamics of superdense plasmas in the ultramegabar dynamic pressure range calls for further investigation. Figure 2.17 shows the pressures currently attained under controllable conditions with the help of shock waves and diamond anvils. One can see that passing to pressures above 10 Mbar calls for the application of unconventional techniques of shock wave generation (subsection 2.2.6), primarily laser-based techniques (section 2.3, lecture 3).

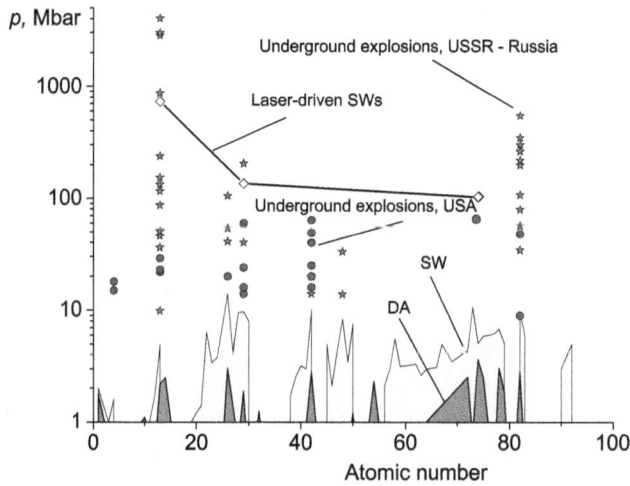

Figure 2.17. Characteristic pressures realized in terrestrial experiments. DA—static technique involving diamond anvils, SW—shock waves driven by light-gas guns and chemical explosives. Reprinted from [18] by permission from Springer. Copyright 2011.

The reader interested in ultrahigh-pressure problems is referred to papers [2, 3, 14, 20, 21].

2.2.5 Explosive-driven magnetic generators

The highest magnetic fields that are possible to obtain in macroscopic volumes on the Earth are produced using explosive magnetic (magnetic cumulative) generators [25–27]. These explosive devices now exhibit record-high magnetic field strengths of 28 MGs and pulsed electric currents of about 300 MA, which corresponds to an extremely high electromagnetic energy density $H^2/8\pi \approx 3$ MJ cm^{-3}. Explosive magnetic generators are the highest-power energy devices today. Their power ranges up to $\approx$100 GW.

A schematic representation of the first explosive magnetic generator proposed by academician A D Sakharov is shown in figure 2.18. In a radial generator (field generator), the initial magnetic flux of induction B_0 is radially compressed by a metal cylinder, which is driven to the center by the detonation of a condensed explosive. With retention of magnetic flux $S = H_0\pi R^2 = H\pi R^2$, the magnetic field strength $H = H_0 \left(\frac{R_0}{R}\right)^2$ reaches many megagauss.

There exist two main limitations [26] imposed on the rate of magnetic flux compression. First, this compression must be rapid enough so as to satisfy the condition $dL/dt \gg R$ and prevent the load damage by the action of ponderomotive forces. Second, since a fast variation of the flux Φ gives rise to the high electric voltage $U = -L dI/dt$, it is necessary to provide a reliable electric insulation against electrical breakdowns.

To date, a family of disk explosive magnetic generators (DEMGs) has been made with HE charges ranging from 240 mm to 1000 mm in diameter. The following parameters were achieved in experiments with DEMGs: an energy gain of 10–30, a

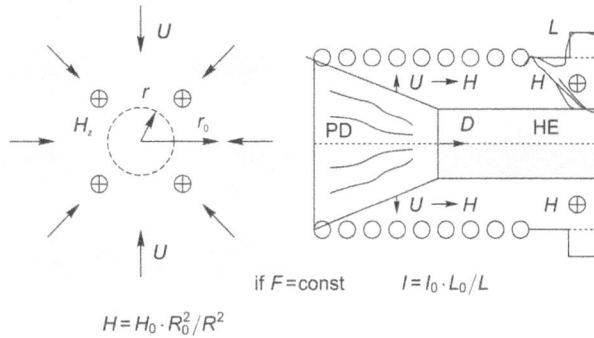

$$H = H_0 \cdot R_0^2 / R^2$$

Figure 2.18. Disk explosive magnetic generator [26]. Reprinted from [18] by permission from Springer. Copyright 2011.

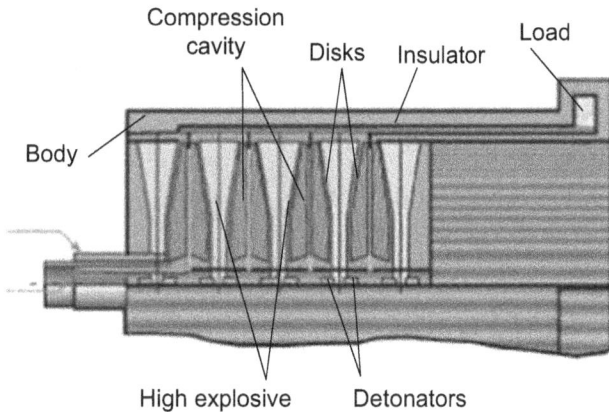

Figure 2.19. Disk magnetic explosion generator [26]. Reprinted from [18] by permission from Springer. Copyright 2011.

characteristic time of 3–10 μs, a specific energy density of 600 J cm^{-3}, and an output energy of 200 MJ. These generators are used to accelerate metal liners to generate shock waves. Velocities of 50 and 15 km s^{-1} were achieved for 1 g and 0.25 kg liners, respectively; moreover, the possibility of quasi-spherical collapse of the liner under the action of an axially symmetric magnetic field was experimentally confirmed.

Figure 2.19 shows the schematic representation of a DEMG. When the magnetic flux in the generator reaches a prescribed magnitude, the generator circuit closes, thereby trapping the introduced magnetic flux. At the same time, the HE charges are synchronously detonated by the initiation system along the axis. Under the action of explosion products, the conducting plates collapse to compress the magnetic flux simultaneously in all cavities and force it out from the compression cavities into a load via a transmission line. The shape of conducting plates is selected so that the compression obeys an exponential law.

Helical explosive magnetic generators (HEMGs) [26], which have a substantially higher inductance and rate of its reduction in comparison with those in other types

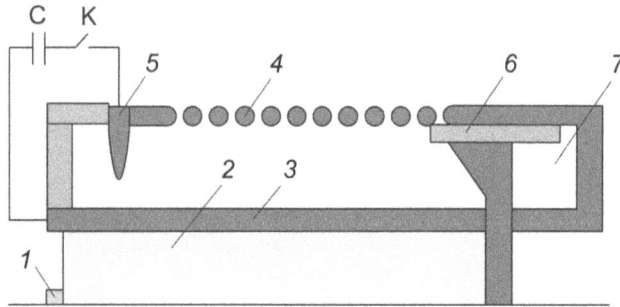

Figure 2.20. Helical explosive magnetic generator: 1—electric detonator; 2—HE charge; 3—liner; 4—solenoid (stator); 5—crowbar; 6—insulator; 7—load; C—capacitor, K—discharge gap [26]. Reprinted from [18] by permission from Springer. Copyright 2011.

of explosive magnetic generators, can efficiently operate on loads with a wide range of inductances and resistances. The key elements of HEMGs (figure 2.20) are a cylindrical solenoid and a metal tube with an explosive charge, which are coaxially arranged and connected via a load. The magnetic flux produced in the generator volume by an external energy source is compressed by the central tube, which, during initiation of an explosive charge at the end face opposite to the load, expands by detonation products to form a cone moving at a detonation velocity along the axis of the device. The central tube is made of soft copper or soft aluminum alloys.

One of the highest-power SEMGs developed at the Russian Federal Nuclear Center 'All-Russian Research Institute of Experimental Physics' (RFNC-VNIIEF) is the generator with a 240 mm internal diameter of turns [26]. For an initial energy of ≈40 kJ, in 50–120 nH loads, it provides a current of up to 15 MA and a magnetic energy of up to 8 MJ. The generator characteristics outperform the world level by a factor of 2–3 in specific energy, by a factor of 10–20 in energy gain, and by a factor of about 2 in specific speed.

Helical explosive magnetic generators have found application for acceleration of solid liners to high velocities. Best known is the series of seven experiments *R-Damage* performed jointly by the RFNC-VNIIEF and the Los-Alamos National Laboratory to study dynamic fracture in convergent geometry using an explosive magnetic device as the driver of a cylindrical aluminum liner, which produces an axially symmetric impact on the target under investigation. The aim of experiments was to study the features of initiation and development of spall fracture as well as of damage compacting in extruded aluminum.

One of the directions of inertial thermonuclear fusion is the generation of high-power soft x-ray pulses with an energy of up to 10 MJ in a time of the order of 10 ns and a compression of a thermonuclear target by these pulses. To generate soft x-rays in the Emir Project, use was made of one- and two-stage liners of tungsten wires ≈0.01 mm in diameter powered by an explosive magnetic generator. At the initial stage of research, advantage was taken of spiral 100 and 200 mm diameter explosive magnetic generators. In these experiments, the current pulses had an amplitude of 2.5–5.5 MA and a pulse rise time of 300–400 ns. A soft x-ray yield of up to 180 kJ was recorded; the duration of soft x-ray radiation was 20 ns and its temperature was about 50 eV. In using a 10 element DEMG with 240 mm diameter HE charges, liners exhibited currents at a level of 14 MA with a characteristic rise time of 1.1 μs. The soft x-ray energy was approximately 0.8 MJ. This is the highest-power soft x-ray radiation source in Russia. It is planned to use in the future DEMGs with explosive charges 480 mm in diameter. Currents in the liner are expected to reach up to ≈50 MA and have a rise time of 0.5 μs. According to calculations, the soft x-ray energy will exceed 10 MJ for these parameters.

Physical experiments using explosive magnetic generators are discussed in detail in [27].

2.2.6 Devices of high-current pulsed electronics

Devices of high-current (10^5–10^7 A) pulsed electronics are employed to produce high-energy-density plasmas in various experimental facilities. Electric energy may have an effect on the direct pulsed Joule heating (electroexplosion) of conductors or the magnetohydrodynamic compression and heating of plasma objects. The stored energy may be used to produce intense bursts of soft x-ray radiation (with a radiation temperature of 200–300 eV) with the subsequent generation of intense shock or radiative thermal waves by this radiation, or for the electrodynamic generation of shock waves, as well as for the electrodynamic acceleration of metal liners. The energy capabilities of such electrodynamic devices are, as a rule (with the exception of the NIF and LMJ), several orders of magnitude higher than for lasers, making it possible to conduct experiments with thicker targets, increase the accuracy of measurements, and relax the time-resolution requirements (10^{-8}–10^{-7} s) to the diagnostic tools.

The electroexplosion of conductors and metal foils by a pulsed current of the order of 50–200 kA is the traditional line of research into the thermophysical properties of refractory materials in the region of condensed matter [28] for characteristic energy densities of the order of 10 kJ cm^{-3} (figure 2.21). This range has been recently extended to 20–30 kJ cm^{-3} with the attainment of strongly supercritical metal states, which allows, in particular, the 'metal–dielectric' transition to be studied in the continuous supercritical expansion of metal plasmas.

The highest plasma parameters have been obtained in high-power Z-pinches of the terawatt power range, in which the electric energy of capacitors, after the corresponding peaking, induces the electrodynamic plasma acceleration followed by the focusing of its kinetic energy on the cylinder axis. Thus, an approximately 10 ns-long burst of 150–200 eV soft x-ray radiation with an energy of ≈1.8 MJ and a power of ≈230 TW was produced in the Z-pinch facility at the Sandia National Laboratories, USA. In these experiments a cylindrical plasma shell was produced by the electric explosion of hundreds of thin (6–50 μm) tungsten conductors by a 20 MA current with a rise time of about 100 ns. A tungsten plasma with an ion density of ≈10^{20} cm^{-3} and a degree of ionization higher than 50 was obtained in the on-axis radiative collapse.

The second interesting application of this facility involves the electrodynamic generation of high-power shock waves. To this end, a high-power electric current pulse

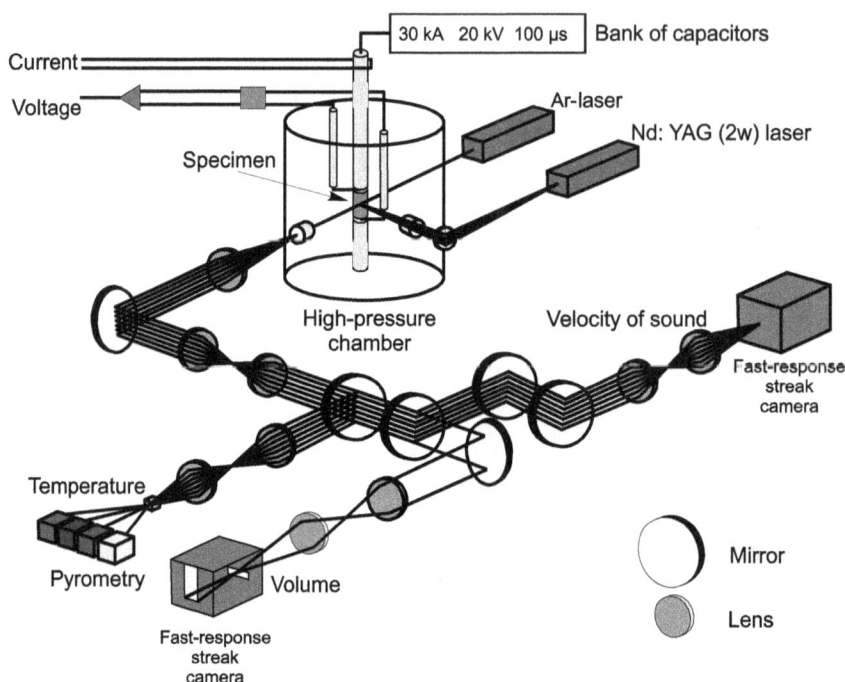

Figure 2.21. Schematic representation of an exploding-wire experiment [28]. Reprinted from [18] by permission from Springer. Copyright 2011.

affected the electrodynamic acceleration of a metal liner to velocities of ≈ 20 km s^{-1} and the generation of megabar-range shock-wave pressures at its impact on a target.

In this case, there is the possibility of controlling the current parameters and realizing a collisionless ('soft') compression of the target to a pressure of about 3 Mbar, with temperatures and entropies of compression lower than in shock-wave heating.

In experiments with the ANGARA facility (figure 2.22), a pulsed current of 4 MA accelerated a xenon plasma liner to a velocity of 500 km s^{-1}. The highly symmetric impact of this liner on the surface of a cylindrical highly porous target excited in it a thermal radiative wave, which emitted soft x-ray radiation with a temperature of about 100 eV. This high-intensity x-ray radiation from the cylindrical cavity was used for highly symmetric generation of plane shock waves with a pressure amplitude of ≈ 5 Mbar, for the excitation of thermal radiative waves with a propagation velocity of 100 km s^{-1}, as well as for the acceleration of metal liners to velocities of 10–12 km s^{-1}.

Interesting plasma parameters were obtained in Z-pinches with an initially gaseous shell of centimeter size, which is ionized by 100–200 ns-long megampere current and collapses to a millimeter-sized column, increasing the plasma density by factors of 20–50 (to 10^{20} cm^{-3}) for an electron temperature of hundreds of electronvolts.

A high-energy plasma is generated in the X-pinch geometry produced by crossing two current-carrying wires.

Figure 2.22. Pulsed ANGARA-5 generator, TRINITI, Russia, intended for controlled thermonuclear fusion and experiments on high-energy-density plasmas and shock and thermal waves. Reprinted from [18] by permission from Springer. Copyright 2011.

Devices of this kind are promising x-ray radiation sources for microlithography and other applications.

In the operation of modern high-current pinches, in the course of production of high energy densities, developed magnetohydrodynamic flows arise, where a crucial role is played by radiation, which is of significance in its own in radiative gas-dynamic simulations of astrophysical objects.

The line of research on the generation of high-energy-density plasmas, based on the quasi-adiabatic cylindrical compression of plasma structures, in which closed magnetic configurations, so-called structures with an 'inverted' magnetic field, are initially created, hold promise. This plasma is additionally compressed and heated by an external heavy metal liner electrodynamically accelerated by a megampere current. Estimates show [11] that the 30 MA pulsed current of the ATLAS facility is capable of producing a several centimeter-long plasma filament 1 cm in diameter with a plasma pressure of about 1 Mbar, an ion density of 10^{19} cm^{-3}, and a temperature of the order of 10 keV.

In the MAGO (MAGnitnoe Obzhatie) project (figure 2.23) [26], a pulsed current of 7 MA delivered by two helical explosive magnetic generators compresses and heats the preliminarily prepared magnetized plasma to parameters close to thermo-nuclear conditions, $\rho \approx 20$ g cm^{-3}, $T \approx 3$–4 keV, which gives a deuterium–tritium (DT) reaction neutron yield of (3–5) $\times 10^{13}$.

In this case, an energy density of $\approx 10^{7}$ J cm^{-3} was realized in the metal liner for plasma compression.

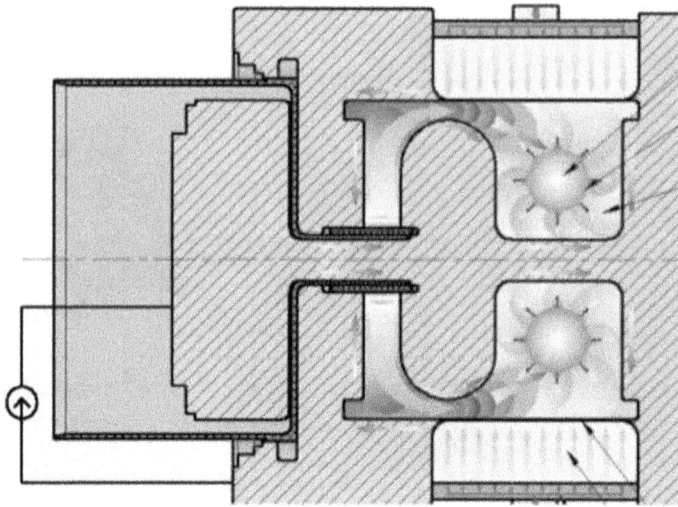

Figure 2.23. Schematic representation of the MAGO plasma generator.

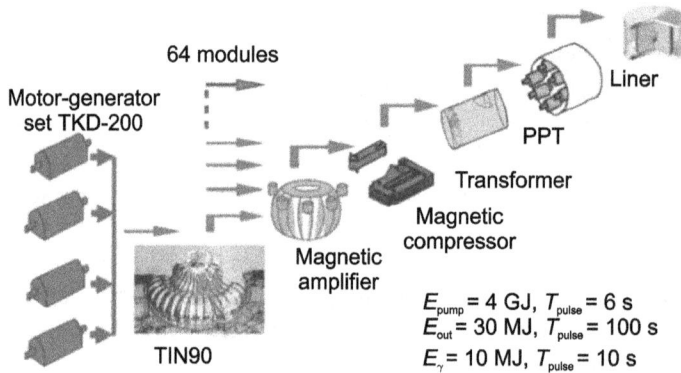

Figure 2.24. Schematic representation of the BAIKAL facility. Reprinted from [18] by permission from Springer. Copyright 2011.

The above-considered devices of high-current pulsed electronics have interesting prospects both for increasing the main parameters, and for developing the elements and schemes of power compression.

The BAIKAL facility projected in Russia should become a source of soft x-ray radiation with an output energy of about 10 MJ intended for experiments with indirect-drive thermonuclear targets and other problems of high-energy-density physics (figure 2.24). In this project, 4 GJ of electric energy is stored in inductive energy storage capacitors and, after the corresponding power peaking, will feed a fast liner system with a power of up to 500–1000 TW.

The success obtained at the Z accelerator (Sandia, USA) is expected to be developed on an X-1 facility feeding two pinch assemblies, each generating 7 MJ of soft x-ray radiation with a power of the order of 1000 TW, which is delivered to a thermonuclear target 4 mm in diameter subjected to x-ray radiation with a temperature of higher than 225 eV for 10 ns. The thermonuclear energy yield of the X-1 facility is expected to be 200–1000 MJ.

A development of the Z accelerator is the IFE-Z Project, which is in fact a prototype of a pulsed thermonuclear reactor. Use will be made of a double Z-pinch, each pinch generating 8 MJ of x-ray radiation at a 66 MA pulsed current. In this case, the thermonuclear energy yield and the gain are planned to be 3 GJ and ≈83, respectively.

The scaling of the dependence of the energy yield of x-rays on the pinch current is shown in figure 2.25.

2.3 High-power lasers

Since the advent of the first laser, one of the most important goals of quantum electronics has been and is still an increase in the peak power of laser radiation [29]. The very notion of a high peak power is permanently changing, and today means a power of no less than 1 PW (10^{15} W). The rapid progress of output laser power in the 1960s and 1970s was based on the principles of Q-switching and mode locking, which has enabled the laser pulse duration to be reduced from microseconds to picoseconds over 40 years. Further advancement in this direction was limited by the large dimensions and cost of lasers and the need to operate at the limit of radiation resistance of optical elements.

The present-day 'renaissance' in laser physics is due to the invention of the chirped (frequency-dispersed) pulse amplification technique (figures 2.26 and 2.27) in 1985, which opened up the way for multiterawatt, petawatt, and even exawatt laser systems [29] to raise the maximum power densities on target to $q \approx 10^{22}$ W cm^{-2}, with the theoretical limit equal to $\approx 3 \times 10^{23}$ W cm^{-2}.

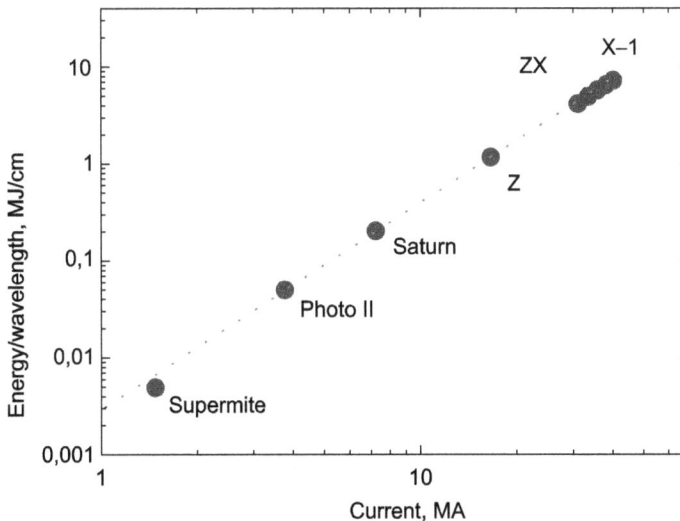

Figure 2.25. Dependence of the soft x-ray yield on the Z-pinch current.

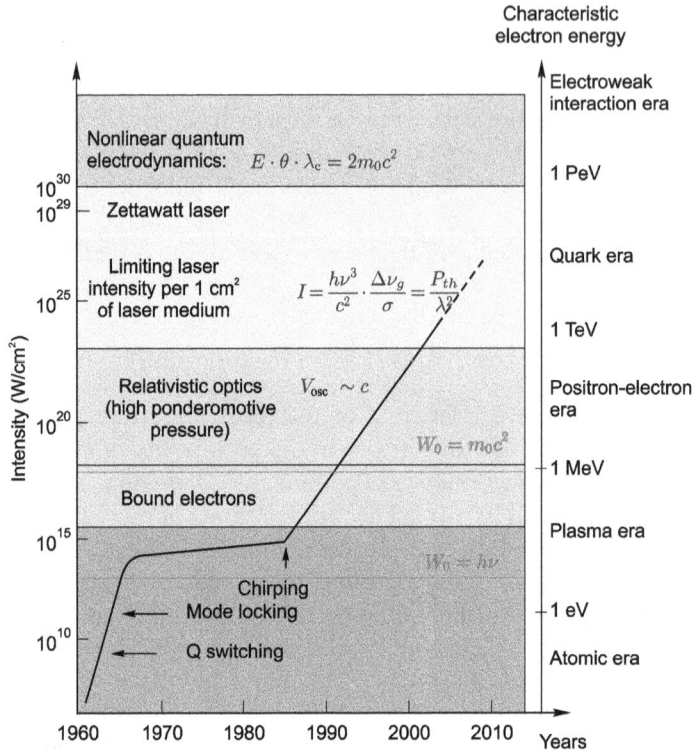

Figure 2.26. Growth of laser radiation intensity with time [11].

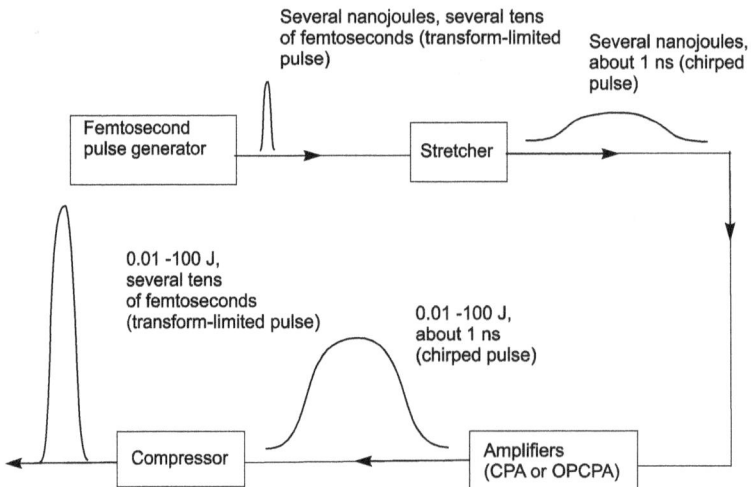

Figure 2.27. General schematic representation of femtosecond lasers.

In this technique (figure 2.27, [30]), an initially short laser pulse is stretched in time, passing through dispersive elements. The pulse is decomposed into spectral components, each of which travels a somewhat different path depending on its wavelength, stretching in time and space several tens of thousands of times due to the separation of its spectral components. Spectral clipping ('chirping') of this time-stretched pulse also occurs, i.e. the frequency continuously varies from the beginning to the end of the pulse. The stretched pulse, having a lower power density, is amplified in the ordinary way by a laser active medium, which now operates under conditions of a much lower flux power, and enters another nonlinear element for optical compression in another dispersion system—the compressor, which is the most energetically stressed element. The peculiarity of such a scheme is that the laser medium amplifies the stretched pulse with a lower intensity. While conventional techniques enabled the radiation to be focused by lenses in two mutually perpendicular directions, the new technique does this simultaneously in three dimensions and sharply increases the resultant power intensity on the target.

Chirping is employed without exception in all lasers with a power of 1 TW and higher [30]. The chirp principle has made it possible to increase the intensity of laser radiation by five–six orders of magnitude and to dramatically lower the cost and dimensions of lasers, which have become 'desktop' devices affordable for small university laboratories. Furthermore, these lasers combine well with big facilities for controlled laser fusion ('fast' ignition, see subsection 2.5.2) and charged-particle accelerators (subsection 2.5.4, figure 2.37) and have provided a way of recording nonlinear quantum-electrodynamic effects such as pair production in vacuum as well as intense optical radiation for studying photon–photon collisions.

Synchronously with the development of the chirp technique, a new method for obtaining superhigh-power pulses was elaborated, based on optical parametric chirped pulse amplification (OPCPA) in nonlinear optical crystals [31]. An advantage of parametric amplification is an unrivaled high chirped-pulse gain: an energy gain of up to three–four orders of magnitude in one pass through the crystal. Another positive point is that the technology of growing wide-aperture crystals of the KDP family (potassium dihydrogen phosphate, KH_2PO_4) has been thoroughly elaborated, which allows the energy of the generated pulses to be increased by scaling the amplifying stages.

Existing and projected petawatt lasers are divided into three types with respect to an amplifying medium: neodymium glass, sapphire, and KDP and DKDP (deuterated KDP, KD_2PO_4) crystal parametric amplifiers (see table 2.3 borrowed from [30]). In all three laser types the energy (in the form of population inversion) is stored in neodymium ions in glass. In the first case, this energy is directly converted into the energy of a chirped pulse, which subsequently undergoes compression. In the second and third cases, the stored energy is converted into the energy of a narrow-band nanosecond pulse, which is then converted into the second harmonic to serve a pump for chirped-pulse amplifiers. This pump either ensures population inversion in a sapphire crystal or decays parametrically into two chirped pulses in a nonlinear crystal.

Table 2.3. The concepts of petawatt lasers.

Amplifying medium Energy source	Nd:glass Nd:glass		Ti:sapphire Nd:glass		DKDP Nd:glass		Cr:YAG ceramics Nd:glass	
Pump		(+)	2ω Nd[a]	(−)	2ω Nd	(−)	1ω Nd[b]	(0)
Pump duration, ns		(+)	>10	(0)	1	(−)	>10	(0)
Amplifier aperture, cm	40	(0)	8	(−)	40	(0)	>50	(+)
Minimal duration, fs	250	(−)	25	(+)	25	(+)	25	(+)
Efficiency (1ω Nd → ϕc), %	80	(+)	15	(0)	10	(−)	25	(0)
Number of petawatt out of 1 kJ 1ω Nd	3.2(3)[c]		6(1.5)[d]		4		10	
Power reached, PW	1.36		1.1		1.0		—	

Note:

Symbols '+', '−', and '0' are indicative of above-average, below-average, and average characteristics, respectively.
[a] Second harmonic of a neodymium laser.
[b] Fundamental harmonic of a neodymium laser.
[c] From the pulse of the fundamental harmonic of a neodymium laser to a femtosecond pulse.
[d] The radiation resistance of diffraction gratings and sapphire crystals limits the peak power at levels of 3 PW and 1.5 PW, respectively.

Unlike neodymium glass lasers, sapphire lasers provide broadband amplification, which allows pulses to be compressed down to 10–20 fs. At the same time, the aperture of sapphire crystals does not exceed 10 cm for the existing crystal growth technology. In an attempt to cross the petawatt threshold, this small aperture will limit the chirped-pulse energy due to optical breakdown and self-focusing.

Table 2.4 and figure 2.28 present the characteristics of laser devices under construction or projected, the commissioning of which will significantly expand the possibilities of experiments in laser plasma physics.

Now in the leading laboratories of the world there are about two dozen installations with a peak radiation power of more than 100 TW and a pulse duration of less than 1 ps [29]. At least ten facilities of the same level are under construction or being upgraded. These are the biggest complexes for laser thermo-nuclear fusion: National Ignition Facility (NIF) (USA), Laser MegaJoule/PETawatt Aquitaine Laser (LMJ/PETAL) (France), and the High Power Laser Energy Research (HiPER) project (Great Britain), which make use (or plan to use) nanosecond laser pulses of tens of radiation channels with a total power of hundreds of TW. Planned for achieving fast ignition (see subsection 2.5.2) is the construction of picosecond multipetawatt laser channels.

The NIF laser system, designed at the Livermore National Laboratory to implement a controlled thermonuclear reaction in the inertial confinement mode, reigns supreme [5, 32]. The NIF comprises 192 laser beams with a total power of 1.8 MJ; focusing on a microtarget must result in its thermonuclear explosion with an energy efficiency ratio of 10–30. The corresponding experiments will be considered in subsection 2.5.2.

Table 2.4. Parameters of laser facilities for high-energy-density physics.

Facility 1	Parameters 2	Year of commissioning 3	Comment 4
		Great Britain	
VULCAN CLF, Rutherford	2.6 kJ, several nanoseconds, 100 TW, 0.5–1.0 ps, 10^{21} W cm^{-2}	1999	Largest European center, which simultaneously has nano- and femtosecond high-power lasers
VULCAN-PW CLF, Rutherford	1 PW, 0.5–1.0 ps	2002	Two-beam, 100–1000 times faster laser, 10 times higher intensity on the target
ASTRA-GEMINI CLF, Rutherford	15 J, 30 fs, one shot per minute, 0.5 PW in each of the two beams	2007	Ultra-intense laser with ultrashort pulse duration and wavelength tunable from IR to VUV, combined with a synchrotron as a diagnostic complex
VULCAN-10PW CLF, Rutherford	10 PW, 0.5–1.0 ps, 10^{23} W cm^{-2}	2014	Ultra-intense laser with ultrashort pulse duration. Use of optical parametric principles makes it possible to obtain a very high contrast of laser radiation on the target
Orion Atomic Weapons Establishment	500 J, 1 ns, 3 ω_0, 10 beams + 2 beams, 500 J, 500 fs, 1 PW, 10^{21} W cm^{-2}		
		European project	
HiPER Future European Facility	40 beams with a total energy of 200 kJ + 24 beams with a total energy of 70 kJ and a duration of about 10 ps for igniting a target	2010	The possibility of operation in the frequency mode is considered

(Continued)

LMJ CESTA, Bordeaux	France	2010–12	2 MJ, 0.35 μm, about 10 ns, 550 TW, 240 beams	The thermonuclear fusion energy yield is expected to reach 25 MJ with an absorbed laser energy of 1.4 MJ
LIL CESTA, Bordeaux		2006	about 40 kJ, 0.35 μm, 9 ns, 8 beams	The first beam was launched in 2003
PETAL Region Aquitaine, Bordeaux		2007–08	3.5 kJ, 0.5–5 ps	An ultra-intense laser works in combination with eight beams of a LIL laser (CESTA), which is a prototype of LMJ
LOA Laboratoire d'Optique Appliquée		2004	30 J, 30 fs, 100 TW, 1–10 Hz	Electrons are accelerated to energies of 200 MeV, beams of monochromatic, collimated high-current electrons are generated
	USA			
NIF LLNL, Livermore		2010	4.2 MJ, 192 beams, 5–25 ns, 1 ω_0	
OMEGA LLE Rochester		1995	30 kJ, several nanoseconds, 60 TW, 60 beams, 3 ω_0	A record number (10^{14}) of neutrons was obtained
OMEGA EP LLE Rochester		2007	6.5 kJ, 3 ω_0, four beams, there will be two beams (High Energy Petawatt-HEPW) with 2.6 kJ, 10 ps in 1 ω_0	Possibility of carrying out cryogenic experiments
	Japan			
FIREX I project Laser LFEX Institute of Laser Engineering, Osaka		2007–08	3.6 kJ in a beam, 10 ps, four beams, total energy of 14.4 kJ	The purpose is to demonstrate that the target can be heated to a temperature of 5–10 keV

FIREX II project Institute of Laser Engineering, Osaka	after 2012	100 kJ, 1 Hz, laser for heating and 100 kJ, 1 Hz, laser for ignition	The power of the thermonuclear reaction and the energy gain are expected to reach 10 MW and 50, respectively. 40% of the energy will be converted into electrical energy. 2 MW of the total energy with an efficiency of 10% will be used for laser operation
China			
SG-II	2005	6 kJ in 1 ω_0, 1 ns, 3 kJ in 3 ω_0, eight beams. 8 TW power in 1 ω_0 for 100 ps	
SG-IIU	2007	3 ns, 18 kJ in 3 ω_0, + 1 PW laser with picosecond duration and energy of 1–2 kJ + 3 ns, 4.5 kJ, laser for x-ray backlighting	
SG-III	2010	3 ns, 150 kJ in 3 ω_0, 64 beams	
Russia			
VNIIEF + IAP RAS	2006	1 PW (100 J, 100 fs)	An on-target laser intensity of 10^{22} W cm^{-2} achieved
Iskra-6 VNIIEF + IAP RAS	Design	300 kJ, 1–3 ns, 128 beams	

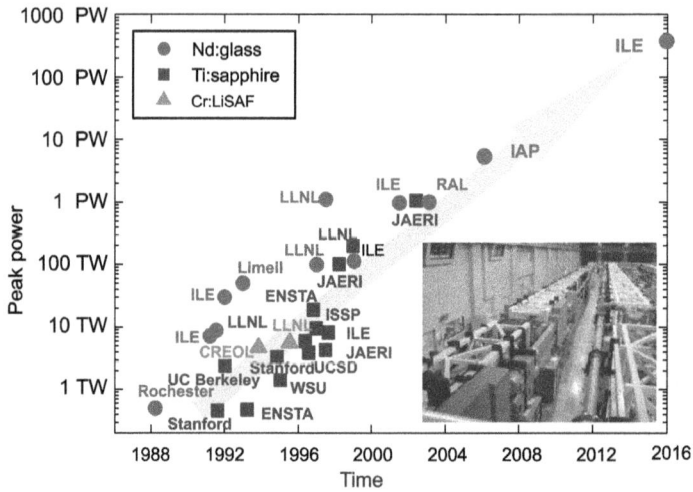

Figure 2.28. Peak powers of laser facilities.

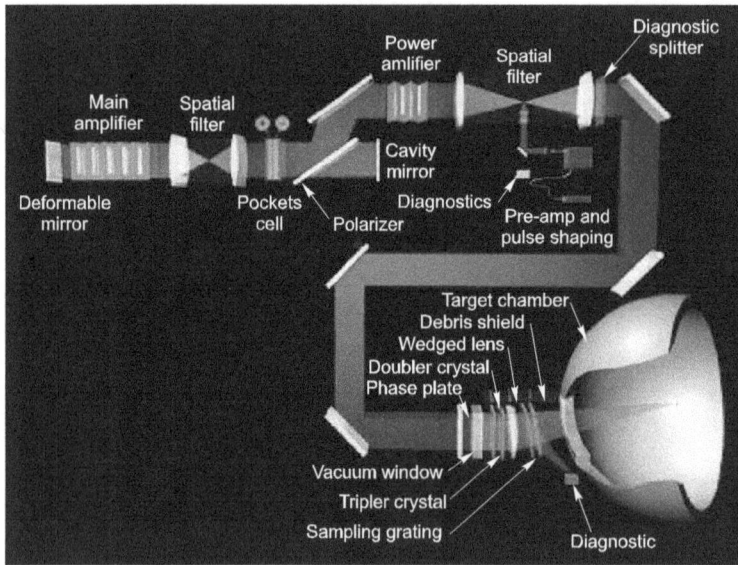

Figure 2.29. Optical scheme of one of 192 beams.

This world's largest laser facility occupies an area of 350 000 m². Each of the 192 beams (figure 2.29 [32]) employs an optical system of 36–38 high-precision large-aperture optical elements and hundreds of smaller-scale optical elements with a surface area of about 3600 m² and a radiation aperture of about 22 m². For comparison, one of world's largest optical reflecting telescopes at the Keck Observatory in Hawaii has an optical surface of ≈152 m², which is only 4% of the NIF.

Figure 2.30. NIF target chamber (USA): a 10 m diameter sphere of 10 cm thick aluminum, coated with a 40 cm thick neutron shield. The chamber weighs about 500 tons. The beam injection system is shown on the right.

A 10 m target chamber (figure 2.30 [32]) provides an input for 192 laser beams and about 100 inputs for diagnostic instrumentation (35 different diagnostic tools). 192 laser beams are focused with high precision onto a thermonuclear target 2.5 mm in diameter (figure 2.54, section 2.5). Each of these beams with an energy of 19 kJ can operate at a fundamental frequency of 1053 nm, the second and third harmonics, and be used not only for thermonuclear fusion, but also for experiments in high-energy-density physics.

The master fiber-optic laser feeds the radiation to a system of fiber-optic elements, which form the desired temporal and frequency pulse shapes. Then, the pulse is branched into 48 channels and each of them is then amplified in preamplifiers, after which it is divided into 192 beams. Each beam depicted in figure 2.29 is amplified in a four-pass amplifier controlled by a wide-aperture Pockels cell. The amplified radiation passes through polarization rotation devices, adaptive mirrors, the target debris shield and is launched into the target chamber. In the construction of the NIF, advantage was taken of the latest achievements of laser technology like high-performance optical glasses, high-performance KDP crystals, etc (see [32] and references therein).

The French LMJ system [6] will radiate an energy of 1.8–2.0 MJ (in the fundamental harmonic) in 240 beams to ensure the conditions for thermonuclear microtarget ignition and will permit modeling different effects of a nuclear explosion. These facilities will make it possible to carry out experiments with shock waves of the gigabar pressure range, thereby advancing to the domain of quasi-classical substance description [10], and to study plasma flows under the conditions of developed radiation effects.

In the RFNC-VNIIEF, an 'Iskra-5' iodine photodissociation facility (figure 2.31) [33] was designed, which consists of 12 laser channels with a total output of 30–40 kJ for a pump pulse duration of 0.36 ns. A capacitive energy storage complex with a

Figure 2.31. Petawatt laser complex at RFNC-VNIIEF.

total stored energy of about 65 MJ was made to pump flashlamps and electric discharge sources.

The target chamber of the 'Iskra-5' facility is equipped with 12 three-component catadioptric lenses. The main 12 channel irradiation experiments were performed with an output energy of 9–10 kJ for a pulse duration of 0.3–0.4 ns. At present, the facility operates at the second harmonic and provides injection of 2.4–3.0 kJ of laser energy into the interaction chamber for a pulse duration of 0.5–0.6 ns. This facility has been used to carry out intense experiments on controlled laser fusion and high-energy-density physics. The next stage—'Iskra-6' facility—will be 10 times higher in power [33].

By 2014, the Rutherford Appleton Laboratory (Great Britain) plans to upgrade the existing 1 PW Vulcan OPCPA laser facility to a power of 10 PW. Two channels of the Vulcan neodymium glass laser, each of energy 600 J, are used to pump two final amplifiers. Advantage is taken of a superbroadband phase matching in a DKDP crystal at a wavelength of chirped pulses near 910 nm. A special feature of this project is a very long (3 ns) chirped pulse.

Two major pan-European laser projects have been recently initiated: HiPER (High Power Laser Energy Research) [34] and ELI (Extreme Light Infrastructure) [35]. The HiPER project is aimed at studying controlled laser fusion at a relatively modest energy of the radiation that compresses the laser target: less than 0.4 MJ in the second harmonic in comparison with 1.8 MJ in the third harmonic on the National Ignition Facility (NIF). This energy 'saving' is achieved by using, along with nanosecond pulses, shorter (about 1 ps) 150–2000 PW pulses to ignite fusion targets (Fast Ignition). The purpose of the ELI project is to design a superhigh-power (50–1000 PW) femtosecond laser to carry out unique research in the field of high-energy-density physics. In these pan-European projects, the femtosecond laser architecture (parametric amplification of chirped laser pulses with a center wavelength of 910 nm in DKDP crystals) is assumed to be optimal for further scaling up.

Petawatt lasers constructed throughout the world will soon become a tool for mastering a new realm of knowledge in high-energy-density physics, i.e. the physics of extreme light fields. In the future, petawatt lasers may be used as charged particle accelerators in basic research, defense technology, and medical applications. Among the last-named ones, mention should be made of the construction of an isotope factory for positron emission tomography as well as of a compact inexpensive ion source for hadron therapy.

These and other potential applications as well as significant progress in the field of petawatt lasers generate interest among commercial companies in mastering the petawatt range, which lends additional impetus to the development of laser technologies. All this gives hope that in 5–10 years petawatt lasers (including OPCPA lasers) will cease to be exotic and become available to many laboratories throughout the world.

To conclude, we note that multipetawatt laser sources capable of generating intensities of 10^{23} W cm^{-2} would be expected to emerge. Today, several countries in the world are simultaneously developing laser facilities with a rated peak power of 10 PW. These are Vulcan-10PW in the UK, ILE-Apollon in France, and PEARL-10 in Russia. Within the framework of the European ELI megaproject, three superhigh-power laser complexes are being built in the Czech Republic, Hungary and Romania, which will be used to study fundamental physics in super-intense fields, problems of generation of attosecond pulses, and photonuclear processes. Anticipated is the generation of monoenergetic electron beams with energies of several GeV, ion beams with an energy at a level of 1 GeV, ultrabright gamma-ray radiation with photon energies of the order of several GeV, ultrashort pulses of subattosecond duration, as well as attainment of record-high radiation intensities at a level of 10^{26} W cm^{-2} with the use of attosecond pulses [29].

2.4 Charged-particle accelerators

The study of the basic properties and structure of matter and the evolution of the Universe since its inception have always been among the key issues of modern physics. In this connection, the generation of extreme states of matter in laboratory conditions along with observations of astrophysical objects has given and continues to give researchers new experimental data on the surrounding matter and form our basic worldviews. A special place among research facilities is occupied by relativistic accelerators, such as the Large Hadron Collider (LHC) (see figure 4.1) at CERN, the Relativistic Heavy Ion Collider (RHIC) (see figure 4.2) at the Brookhaven National Laboratory (BNL), USA, as well as the cyclotron complex at RIKEN (Japan) and TWAC-ITEP and Nuclotron accelerators at the JINR (Russia).

In the next three subsections, we will briefly consider the projects of fundamental devices that are being implemented or ready to be implemented, which are expected to make a significant breakthrough in the near future.

2.4.1 Large Hadron Collider

Following [36] we begin our consideration with the Large Hadron Collider—a facility with the highest kinetic energy of colliding particles. It gives experimenters a possibility to reach a new—teraelectronvolt—energy level and study the properties of space on a smaller scale. We are dealing with verification of the basic predictions of the Standard Model and the quest for effects beyond its framework.

Of greatest interest is the nature of the emergence and diversity of particle masses and fields, the structure of physical vacuum, particle type multiplicity in the Universe, the unambiguous description of fundamental forces, including gravitation, the possible existence of supersymmetric partners of all observable particles and additional space–time dimensions, as well as several other basic issues.

The Large Hadron Collider operates in two modes. In proton mode, beams of colliding protons are accelerated to 17 TeV (the energy of each colliding proton beam is equal to 7 TeV) in the center of mass system (in the rest system of one of the protons, this is equivalent to 10^{17} eV). In the second, heavy-ion mode, the specific energy of ions is about 5.5 TeV/nucleon.

The oppositely directed particle motion in two channels is ensured by 1232 superconducting magnets with an induction of 9 T each (figure 2.32). The force acting on each magnet is approximately 100 tons.

The magnets are cooled to very low temperatures of 1.9 K and pressures of 10^{-10} Torr. To initially cool the magnets, use is made of up to 1.2×10^7 l of liquid nitrogen and up to 7×10^5 l of liquid helium for their subsequent operation. Engineers have to provide reliable control over 40 000 connections to prevent them from leaking.

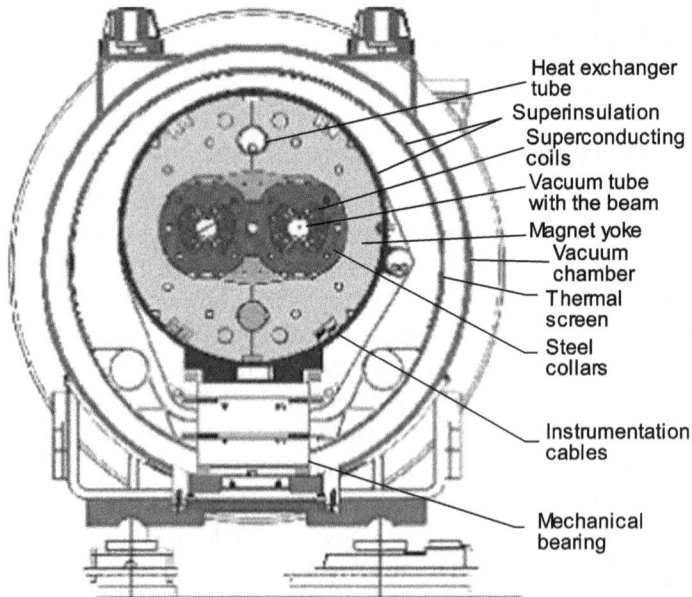

Figure 2.32. Schematic representation of the LHC magnetic system.

Apart from the principal magnets, there are over 500 quadrupole superconducting magnets and over 4000 beam-positioning superconducting magnets.

Apart from the high energy of colliding particles, the LHC accelerator has a high luminosity of up to 10^{34} cm^{-2} s^{-1}, which makes it possible to study rare events with small cross sections. These rare events are of interest for the new Physics for the sake of which the LHC was constructed.

To study processes with large cross sections, it is sufficient to operate in a lower-luminosity mode. This may be attained in different ways, for instance, by beam defocusing, like in the operation with the LHCb (Large Hadron Collider beauty experiment) detector.

Four main detectors—ATLAS (A Toroidal LHC Apparatus), CMS (Compact Muon Solenoid), ALICE (A Large Ion Collider Experiment), and LHCb—register products produced at four points of beam interaction. To this end, four international collaborations with the same names were developed for operating these detectors (figure 2.33).

The largest-scale detectors, ATLAS and CMS (figures 2.34 and 2.35), are intended for studying proton–proton and ion–ion collision events. The ALICE detector is designed to investigate ion–ion collisions, although some data will also be obtained for proton–proton processes, and LHCb is intended for recording events with the b-quark production. In addition to the four main detectors, there are also smaller subdetectors that work in combination with the main detectors: LHCf (LHC forward), Totem, ZDC (Zero Degree Calorimeter), FMS (Forward muon spectrometer), Castor, and FP420.

The detectors are outstanding engineering structures. Thus, the height of the ATLAS detector is 25 m (which is approximately equal to the height of an eight-storey building), its length is 46 m, the weight is 7000 ton.

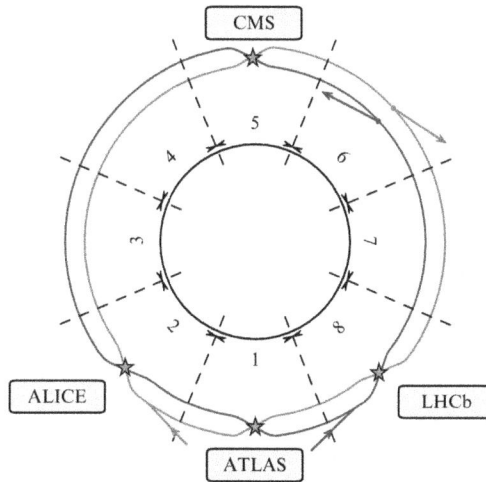

Figure 2.33. Schematic representation of the LHC accelerator: arrows show directions of propagation of beams, and asterisks indicate four points of their collision in which four main detectors are located.

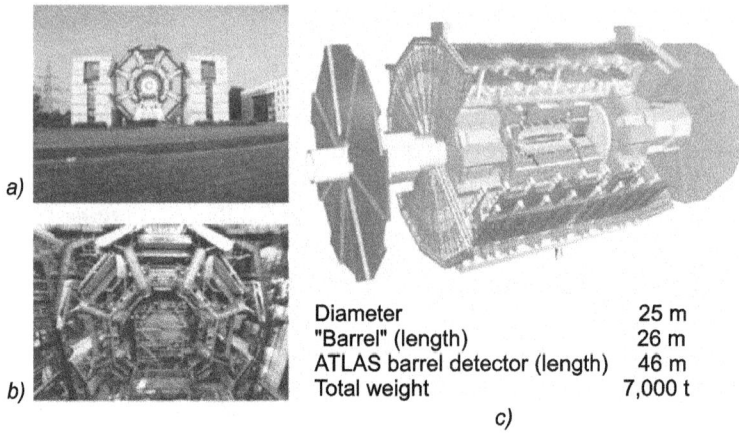

Diameter	25 m
"Barrel" (length)	26 m
ATLAS barrel detector (length)	46 m
Total weight	7,000 t

Figure 2.34. ATLAS detector: (a) life-sized image of the detector superimposed on building 40 at CERN, where the ATLAS and CMS collaboration is sited; (b) general view; (c) cross-sectional scheme.

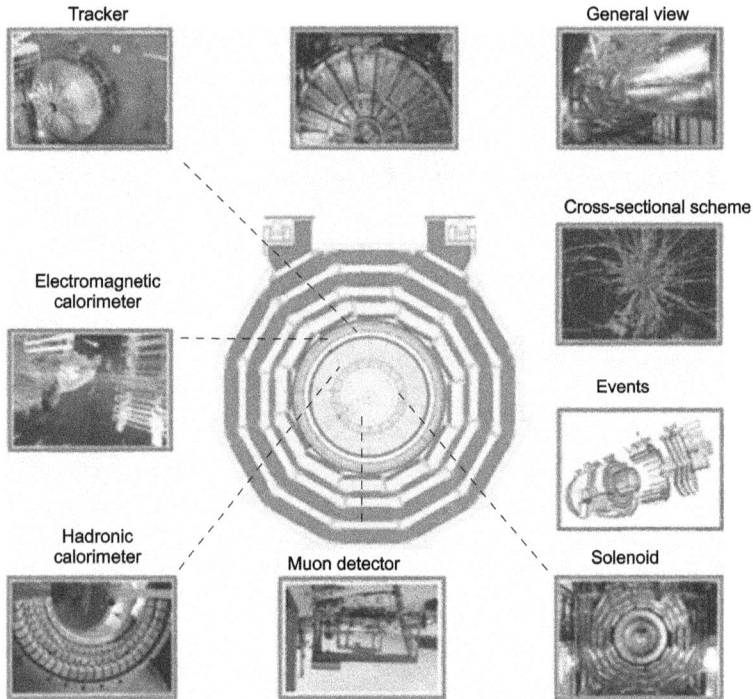

Figure 2.35. CMS detector: cross-sectional scheme and internal structure with indication of trackers and calorimeters.

Such dimensions are needed to accommodate all trackers and calorimeters so that to ensure as large a range of spanning angles as possible in which the generated particles will be recorded.

The structure of the beams has its own peculiarities. The proton flux in the accelerator ring consists of bunches spaced at 25 ns in time, or at a distance of about 7.5 m from each other. At this repetition rate of the beams, the frequency of their encounters at the interaction points is equal to 30–40 MHz. Most protons of the beam will overfly these points without any interaction and will continue their motion along the accelerator ring. But even in this case, about 20–40 proton–proton interaction events will simultaneously occur at a very short range of every beam encounter. The trigger system must distinguish between the interactions and, furthermore, select those of them which are of physical interest. The triggers must therefore be fast-responding, purpose-oriented, and suited to the corresponding detector goals [36].

Magnetic fields in detectors are necessary for bending the trajectories of the produced charged particles and reach 2 T in ATLAS and 4 T in CMS. The trajectory curvature allows the momentum of a particle to be measured and its mass to be determined. This is the main information acquired about the interaction. The detection of particles is based on the interactions of charged particles with the substance of the detector. The general 'architecture' of ATLAS, CMS, and ALICE detectors has much in common, whereas LHCb is oriented to b-quarks traveling mainly in the forward direction.

The LHC project necessitates the use of huge computational resources to store, distribute, and analyze the information arriving from the accelerator. It amounts to several tens of petabytes per year. To this end, a special Grid project is being developed, which unites computer centers around the world.

The main scientific tasks of the LHC project are discussed in [36].

Further research in the field of high-energy-density physics and the construction of new accelerator facilities depend to a large extent on the results obtained at the LHC.

A promising project, known as the SLHC (Super LHC), will improve the statistics and expand the range of masses where Higgs bosons, supersymmetric particles and additional dimensions will be searched for.

Another project is the International Linear Collider (ILC), where electrons and positrons counterpropagating along rectilinear trajectories will collide with a total energy of 1 TeV. It will be possible to study the supersymmetric world with a good precision for large masses. Apparently, there will be a need for even higher energies of several teraelectronvolts. Then the Compact Linear Collider (CLIC) project will become topical, if the need arises to investigate particles of even larger masses.

These projects will compete with the projects which plan to double (DLHC) or even triple (TLHC) the LHC energy [36].

The discussion and comparison of different possibilities of choice are now underway. Concrete solutions can be adopted only after the experimental information from the LHC becomes available. The future of high-energy-density physics largely depends on these results.

2.4.2 FAIR project

The European Facility for Antiproton and Ion Research (FAIR) is the project of a new research complex based on a multipurpose accelerator with antiproton and radioactive-nuclei beam parameters that are unrivaled in the world and which offer unique possibilities for carrying out scientific investigations in the most topical areas of modern science and innovative technologies [37].

The ultramodern accelerator complex, being built in Darmstadt, Germany, will provide researchers with high-energy, precisely tuned beams of antiprotons and different ions with unique quality (brightness, phase density) and intensity. A distinctive feature of the FAIR complex is the presence of high-intensity primary and secondary beams of stable and radioactive nuclei, as well as of beams of antiprotons, which exceed in intensity the existing beams in the world by 100–10 000 times.

In addition, the FAIR experiments will make it possible to make significant progress in the study of unknown regions of the nuclear matter phase diagram in comparison with the experiments on the RHIC ion collider at the Brookhaven National Laboratory, USA, and the LHC collider at CERN [8]. In contrast to these laboratories, where the main attention is paid to the investigation of the properties of nuclear matter at extremely high temperatures but low baryon densities, the FAIR experiments are aimed at a detailed study of the properties of matter at the highest baryon densities attainable under terrestrial conditions.

The FAIR project is an international project involving 15 countries: Austria, Great Britain, Germany, India, Spain, Italy, China, Poland, Russia, Romania, Slovakia, Slovenia, Finland, France, and Sweden. Countries intend to make contributions in the form of deliveries of high-tech equipment and components of experimental installations, as well as in the form of investments.

The total cost of the project, including construction, maintenance and operation for a period until 2025, is about 3 billion euros. The cost of construction of the initial phase of the acceleration center and its basic research facilities exceeds 1.5 billion euros.

The construction period of the FAIR facility is 8 years. It is planned to complete its construction in 2018. At the same time, scientific research on a number of subsystems of the accelerator complex is expected to begin as early as 2017.

To date, 14 large international collaborations have been organized, which are grouped into four subject lines of research:

- Nuclear STructure, Astrophysics and Reactions (NuSTAR);
- Compressed Baryonic Matter (CBM);
- antiProton ANnihilations at DArmstadt (PANDA);
- Atomic Physics, Plasma, and Applied sciences (APPA).

About 3000 researchers from all over the world will carry out experiments on the FAIR accelerator complex, which are aimed at the study of the basic properties and structure of matter and the evolution of the Universe since its inception.

Three basic experiments form the core of the experimental program on electro-magnetic plasma:

- Heavy Ion Heating and EXpansion ('HIHEX' experiment);
- LAboratory PLAnetary Science ('LAPLAS' experiment);
- Warm Dense Matter ('WDM' experiment).

The most important feature of these three experiments on the physics of high energy density in matter on the FAIR accelerators will consist in the unique possibility of plasma diagnostics using high-energy proton beams (radiography) with the simultaneous use of high-power petawatt laser radiation. This laser is capable of generating an intensity of 10^{21} W cm^{-2} on a target. The combination of a high-intensity laser pulse, a diagnostic proton beam, and a high-intensity ion beam opens up new unique experimental possibilities for the FAIR.

HIHEX experiment. High-intensity heavy ion beams with a pulse duration of 50–100 ns make it possible to rapidly (in comparison with the characteristic time of hydrodynamic motion) heat a substance and then observe the expansion of the hot substance in the surrounding medium, i.e. create a high level of energy release and successively observe the isentropic expansion. In an experiment of this type [8], the heated material will go through several new interesting states in its expansion: as a result of heating, the initial metal of normal density will reach a superheated liquid state with a disordered ion component and degenerate electrons. During isentropic expansion, the substance passes the state of a quasi-nonideal Boltzmann plasma and a rarefied gas. In the course of further expansion, the degree of degeneracy decreases, which is accompanied by redistribution of the energy spectrum of ions and atoms, as well as partial recombination of a dense plasma. In a disordered electronic system, metal–dielectric phase transitions can occur, and the plasma near the critical point and the liquid–vapor phase equilibrium becomes nonideal. When the isentrope enters the two-phase liquid–vapor domain, the gas phase starts condensing. At higher levels of energy deposition, the isentropic expansion may be attended by still more exotic effects with a strong variation of the plasma ionization degree α and nonideality parameter Γ. The range of changes in the thermodynamic parameters of a substance in one experiment may vary over a broad range: up to six orders of magnitude in pressure and up to four orders of magnitude in density.

In the pressure–entropy diagram (figure 2.36 [8]), the states of matter with a high energy density, including the hot compressed ionized substance, the nonideal plasma, the hot expanding liquid, and the quasi-ideal plasma, occupy a vast domain.

The presently available information about metal properties obtained by the method of isentropic expansion of a shock-compressed material is concentrated primarily along the Hugoniot adiabat and is supplemented with model estimates of the position of critical points in the phase plane. That is why the vast domain of the phase diagram under the shock adiabat, including the domains with the critical points of metals and the nonideal plasma domain ($\Gamma \geqslant 1$), calls for further investigations [13].

Figure 2.36 shows the phase diagram domains corresponding to parameters attainable with the SIS 18 and SIS 100 heavy ion accelerator facilities. Thus, the SIS

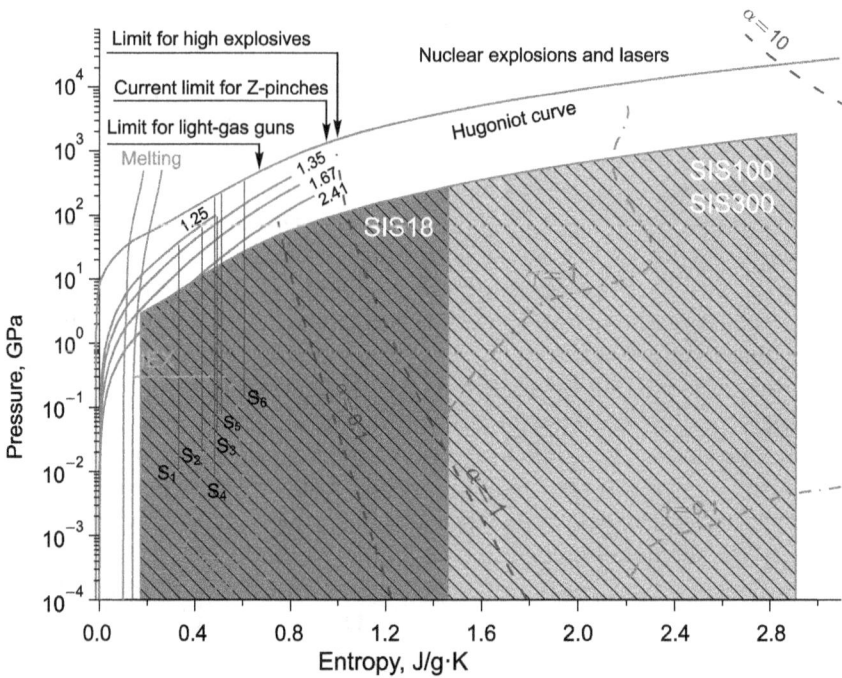

Figure 2.36. Possibilities of SIS 18, SIS 100, SIS 300 heavy ion accelerators (see table 2.2) for the generation of high energy densities in lead.

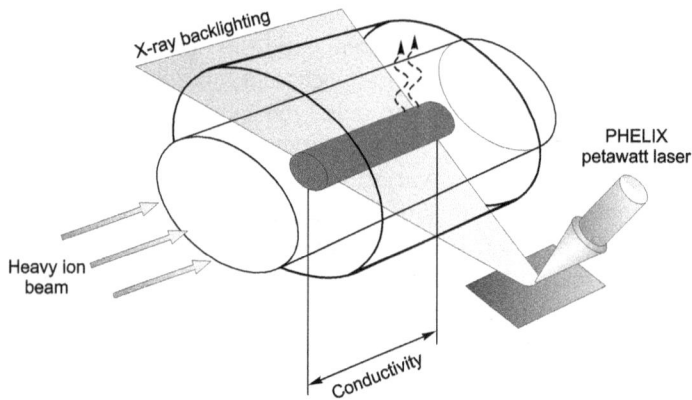

Figure 2.37. Schematic representation of the HIHEX experiment of the FAIR project using a relativistic heavy ion beam and a petawatt laser.

18 operating accelerator produces a heating uranium ion pulse with an intensity of about 10^{10} particles, a duration of about 100 ns, and an ion energy of 300 MeV/nucleon.

When this pulse is focused onto a target, the energy release is at a level of 1 kJ g^{-1}. In the future it is planned to increase the beam intensity and raise the energy release to a level of 10 kJ g^{-1}. The TWAC-ITEP storage accelerator launched at the Institute of Theoretical and Experimental Physics, Moscow, in 2003 is aimed at reaching an energy release at a level of 10–20 kJ g^{-1} by concentrating on a target the pulses of copper or cobalt ions with energies of up to 700 MeV/nucleon. The new SIS 100 accelerator in the FAIR project will provide an energy release of above 100 kJ g^{-1}.

Figure 2.37 shows the schematic representation of the HIHEX experiment, in which a cylindrical target is bulk heated by a high-intensity heavy ion beam [8]. The heated substance (plasma) begins to expand isentropically into the surrounding vacuum. The expansion parameters necessary for the construction of the equation of state of matters are measured using diagnostic systems that include x-ray back-lighting from a petawatt laser.

Two large international collaborations (HEDgeHOB and WDM) aim to use the property of volumetric energy release by an intense ion beam in a substance to generate states of matter characterized by a high level of entropy. The high-power pulse beams generated by the FAIR accelerators will be capable of producing large volumes of dense plasma with an energy density, which is of interest for the solution of a broad range of basic problems in plasma physics, radiation hydrodynamics and magnetohydrodynamics, radiation material science, planetary geophysics, atomic and molecular physics, etc.

The 'LAPLAS' experiment implements the mode of strong substance compression by cylindrical implosion on the cylinder axis. Interest in this phenomenon is due to the feasibility of generating under laboratory condition the planetary states of matter existing in the central parts of giant planets (Jupiter and Saturn), where the substance has a density of 1–2 g cm^{-3} at a pressure of 5–10 Mbar and a temperature of several electronvolts. Furthermore, the experiment is related to the problem of inertial confinement nuclear fusion driven by a heavy ion accelerator [8]. As a rule, the targets are cylindrical structures with layers of different initial density (figure 2.38 [37]).

The central region is occupied by a substance that must be compressed to the maximum attainable final density, for example hydrogen or an equimolar mixture of DT. A layer of substance that absorbs the energy of the ions (absorber) is exposed to a 'tubular' beam of a special shape, which has an annular cross section.

In this case, special emphasis is placed on the selection of the combination of the parameters of the heating ion beam pulse, the geometrical dimensions of target

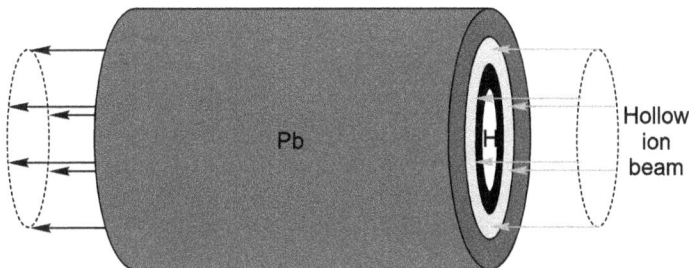

Figure 2.38. Structure of a multilayer cylindrical target for the 'LAPLAS' experiment.

layers, and their initial densities to provide a close-to-adiabatic mode of fuel compression on the cylinder axis—the so-called quasi-entropic mode.

A characteristic feature of the experiment is to provide a mode for the generation of a multitude of successive, reverberating weak shock waves compressing hydrogen along the isentrope. The results of hydrodynamic calculations show that as the hydrogen–lead interface executed a relatively slow adiabatic motion towards the cylinder axis between this interface and the axis, there emerges a series of weak reflected shock waves, which produce a state of matter corresponding to the conditions of the metallization of hydrogen. In accordance with the equation of state from the SESAME table, this is a pressure of 3 Mbar, a density of about 1 g cm^{-3}, and a temperature of less than 0.1 eV.

These parameters remain invariable for 160–200 ns, which is quite sufficient for measuring experimentally the conductivity of hydrogen under extreme conditions. Recent experiments on quasi-adiabatic compression of hydrogen [8, 23] indicate that the pressure-induced metallization sets in even at pressures of about 1 Mbar and a density of about 0.6 g cm^{-3}.

The aim of the **WDM experiment** is the realization of the quasi-isochoric mode of matter heating and achieving the extreme state of the plasma under strong interparticle interaction at large values of the nonideality parameter $\Gamma \geqslant 1$ [23].

The physical substantiation of the forthcoming experiment with ion beams of the SIS 100 accelerator being developed within the framework of the international FAIR project is presented in paper [38]. The experiment is aimed at the study of the solid hydrogen state for an energy deposition level of 130 kJ g^{-1}, which is provided by a 200 MeV/nucleon uranium ion beam with an intensity of 8×10^{10} W cm^{-2} focused to a spot of radius $r_b = 350$ μm. According to the equation of state from the SESAME tables, this corresponds to a temperature of 0.6 eV—the mode of 'warm dense matter', in which the entire beam energy is converted into the internal substance energy.

In the case of volume energy release by an ion beam, typical of ions with energies $E \geqslant 10$ MeV/nucleon, the specific energy deposition E_s is the determining characteristic, which can be measured with good accuracy. Moreover, if the sample's substance density ρ_0 remains invariable during heating, the thermodynamic substance parameters after irradiation are defined by the quantities ρ_0 and E_s. Therefore, any measurable physical quantities are functions of this, well defined, thermodynamic state.

The choice of the target material is determined by the capabilities of the diagnostic technique, which involves recording of the spectral and angular distribution of the x-ray photons scattered by the substance of the heated sample—the Thomson x-ray scattering method. This x-ray backlighting with a temporal resolution can be provided by the PHELIX petawatt laser, which is under construction at the GSI, Darmstadt. For x-ray photon energies of ≈ 1–3 keV, the choice of target structure materials is nevertheless limited to low-Z elements. To carry out the substance state diagnostics and interpret the data, it is advisable to deal with a uniform density distribution over the sample volume. The simplest target for

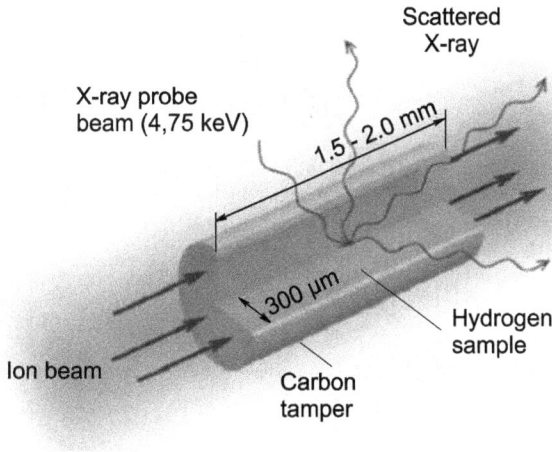

Figure 2.39. Schematic representation of the WDM experiment on quasi-isochoric heating of hydrogen.

a quasi-isochoric experiment is a cylinder of frozen hydrogen of radius $R_h \geqslant r_b$ (figure 2.39 [38]).

For a rectangular intensity distribution over the beam cross section, the density on the cylinder axis remains constant until the unloading wave reaches the axis. However, for a realistic beam with a Gaussian intensity distribution over its cross section, for which the second derivative of pressure with respect to radius is nonzero, the density begins to decrease even prior to the arrival of the unloading wave at the target axis.

This effect of the hydrodynamic unloading of the heated target region may be compensated by using an inertia shell (tamper) to confine the frozen hydrogen. To provide the desired 'confinement' of the heated substance by a low-Z material transparent to x-ray photons, the tamper is also heated by the peripheral part of the ion beam. In this case, the heated layer of the tamper produces counterpressure for confining the main substance of the target material. For a tamper it is evidently advantageous to use a material with a high sublimation energy to delay the onset of the hydrodynamic expansion of the tamper itself.

Numerical simulations performed using the BIG-2 two-dimensional hydrodynamic code demonstrate that the tamper density should be less than that of graphite. That is why a plastic with a density of 1.5 g cm^{-3} under normal conditions was chosen as the tamper material. The temporal behavior of the target layer densities is shown in figure 2.40 [37]. Initially the hydrogen density begins to decrease due to the Gaussian profile of the ion beam. The pressure in the tamper material exceeds that of hydrogen, and therefore the tamper begins to move inwards and generates a weak shock wave. Later on, when the tamper density becomes lower, the pressure equalizes and the hydrogen–tamper interface stops. Then, the increasing hydrogen pressure returns the interface to its initial position. According to the calculations, by the end of the ion irradiation, an almost uniform density is achieved along the target radius.

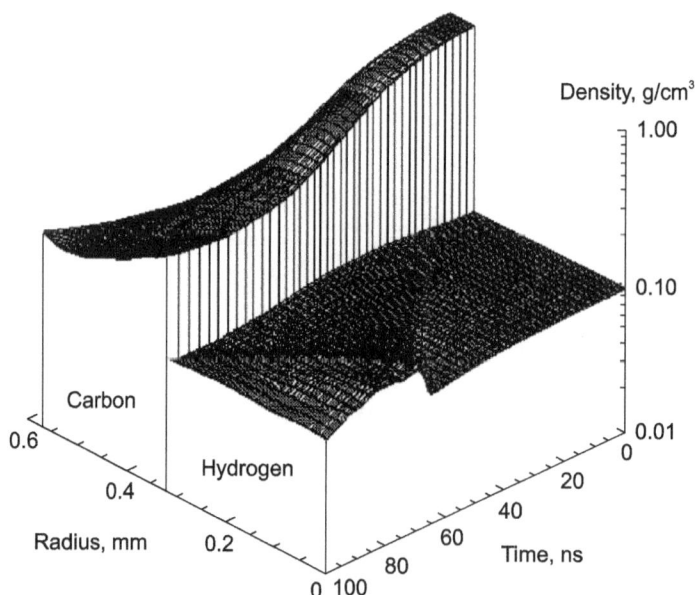

Figure 2.40. Dependence of the density of target layers on time.

The calculations show that an ion beam is capable of ensuring, for a prescribed set of initial parameters, a quasi-isochoric heating mode of solid-density hydrogen for a temperature of 5000 °C, i.e. to provide dynamic confinement within 100 ns.

Radiation material science and biophysics (BIOMAT). Heavy ion beams of the FAIR accelerator complex with energies of 10 GeV/nucleon and higher are of immediate interest for the study of radiation influence of galactic and solar particles on living biological objects as well as on various materials. Areas of applications include radiation material science, radiation resistance of the microelectronic components, nuclear medicine, simulation of the effect of cosmic radiation of heavy charged particles on living cells and materials. In this case, of special significance in the formulation and execution of these experiments is the property of fast ions to penetrate to a controllable depth inside the samples.

Heavy ions generated by the SIS 100 will be used in the multipurpose BIOMAT facility equipped with a raster scanning system, which will provide an excellent beam quality under uniform irradiation of a large-surface field in a broad fluence range of particles with different atomic masses. A special robot will be used for positioning the samples (figure 2.41 [37]).

A broad set of ion beams—from proton to uranium ones—will be required for biophysical and materials research. The highest required beam intensities for an energy of 10 GeV/nucleon in the ion mass range from protons (5×10^{10} cm^{-2}) to iron (1×10^8 cm^{-2}) correspond to a dose of 10 Gy uniformly absorbed by a target of 100 cm^2 area in 1 min Acceptable irradiation times may be achieved for this dose rate, which is highly significant in the irradiation of sensitive biological samples (i.e. cell cultures) that should be in a nutrient medium. The highest intensities of heavier-ion beams—from krypton to uranium ones (1×10^8 cm^{-2})—are determined by the

Figure 2.41. Robotized system for the automatic positioning of biological samples.

fluxes required for materials research in combination with reasonable exposure times.

Heavy ions release energy for a short time in a small volume, providing a threshold change in the structure of the solid, which was in the subthreshold state due to preliminary compression. In addition, the ion beam generates acoustic waves in the sample, which also affect the processes of phase transitions.

Thus, the combination of high pressure and fast energy release by the ion beam in the local region will open up new possibilities in generating phase transformations that are impossible when only one pressure is applied. In these investigations it is planned to employ several promising diagnostic techniques, including x-ray photon and neutron diffraction, optical Raman spectroscopy, and also the resonance measurements of muon spins.

In conclusion, it should be noted that these investigations will be carried out on a unique research complex, which has no analogues in the world. A distinctive feature of the FAIR complex is the capability of generating high-intensity primary and secondary beams of stable and radioactive nuclei as well as of antiproton beams, which exceed in intensity the presently available beams by a factor of 100–10 000 and offer unique possibilities for performing research in the most topical areas of modern science and technology even in the initial version of the project. A high quality of the beams will be provided due to the capability of their storage, cooling, and compression. The proposed scheme of the accelerator complex has the principal possibility of optimizing the parallel self-consistent operation of the complex. In general, the complex will operate for various experiments as if it were intended for each experiment separately.

2.4.3 Heavy-ion experiments in the NICA project

The first accelerator in Russia that allowed beams of high-energy nuclei to be generated was the synchrophasotron of the Joint Institute for Nuclear Research

(JINR) in Dubna, which has been replaced by the nuclotron belonging to a new generation of accelerators based on the use of superconducting magnets. Since the first experiments on the acceleration of relativistic deuterons at the synchrophasotron and the study of the formation of cumulative particles in reactions involving nuclei, which initiated a new direction, called the relativistic nuclear physics [39], several generations of accelerators of ultrarelativistic heavy ions have succeeded each other, and the energy scale has grown in the laboratory equivalent from several gigaelectronvolts per nucleon to several tens of teraelectronvolts per nucleon.

In experiments with heavy ions, a trend towards a steady increase in the energy of the operating accelerators has been observed throughout the entire time. In pursuit of higher energy, more powerful and expensive facilities have been made. Quite recently an opposite tendency has begun to show, namely, the trend for lowering the accelerator energy to obtain evidence for the existence of new states of nuclear matter and to study the corresponding phase transitions and critical phenomena. In this case, a deeper thermalization of the substance occurs in relativistic nuclear collisions and, as a consequence, new forms of nuclear matter become clearly manifested in the experiment. Now it is believed that moderate accelerated ion energies are required for this purpose: 8–40 GeV per nucleon in the laboratory frame of reference.

According to the existing theoretical notions, nuclear matter may undergo a series of phase transitions of the first kind with an increase in temperature and/or baryon charge density. Among them is the phase transition of the first kind involving the recovery of special nuclear interaction symmetry—chiral symmetry, which is severely violated at low temperatures and/or baryon charge densities and is recovered at high ones. As a consequence, the existence of a mixed phase corresponding to this transition is predicted, i.e. the phase of existence of matter with broken and unbroken chiral symmetry. As we saw above, another predicted phase transition of the first kind is a so-called nuclear matter deconfinement transition to a hypothetical quark–gluon plasma state.

The lowering of energy in heavy ion experiments is, in a sense, a 'return to the origins', because this is a return to the previously mastered energy values. However, its expedience is quite evident: a unique possibility thereby appears to take advantage of state-of-the-art detectors, modern notions and models, which were nonexistent at the time of execution of the previous experiments at the same energies, so as to refine the existing views, perhaps revise past errors, as well as to supplement the existing experimental data with new ones required from the viewpoint of modern concepts.

The collider project named Nuclotron-based Ion Collider fAcility (NICA) (figures 2.42 and 2.43), which has been carried out by the JINR in close cooperation with the institutes of the Russian Academy of Sciences, Rosatom, the Ministry of Education and Science of the Russian Federation, and a number of other organizations since 2007, is devoted to exploring relativistic heavy ion collisions at energies $\sqrt{S_{NN}} \approx 4$–11 GeV in a center-of-mass system or at 8–60 GeV in the laboratory frame. This ambitious project intends to investigate the properties of heated and heavily compressed nuclear matter produced in heavy-ion collisions for

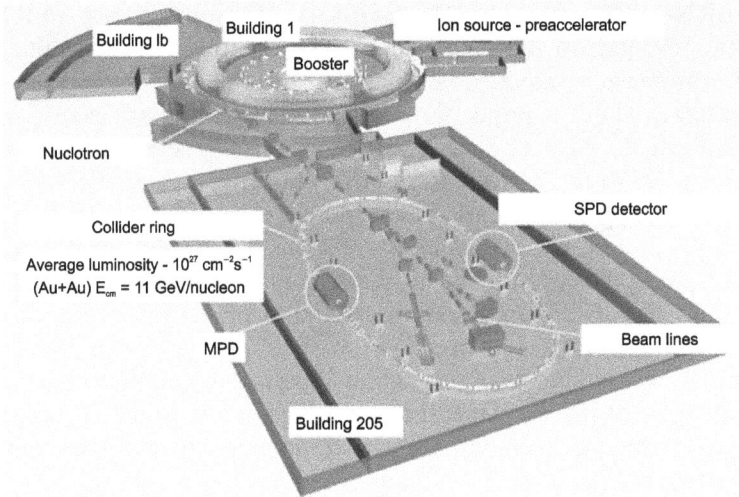

Figure 2.42. Location of the NICA collider in the JINR accelerator complex area [40].

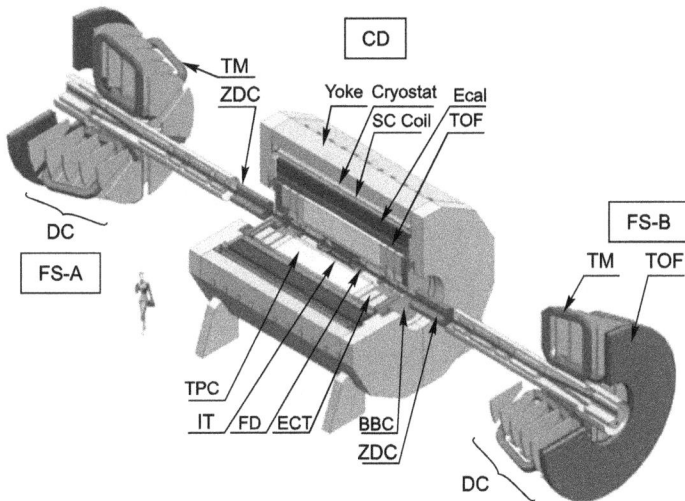

Figure 2.43. Schematic representation of the multipurpose MPD detector [central detector (CD) and two forward spectrometers (FS-A and FS-B)]: IT—internal tracker; TPC—time-projection drift chamber; TOF—time-of-flight counters; Ecal—electromagnetic calorimeter; ECT—end-cap tracker; ZDC—zero-degree calorimeters; TM—toroidal magnets; DC—drift chambers. Reprinted from [18] by permission from Springer. Copyright 2011.

the highest baryon charge densities attainable under laboratory conditions, and to search for phase transitions, new nuclear matter states, and new manifestations of the production of a mixed quark–hadron phase.

The average luminosity for $Au^{79+} + Au^{79+}$ collisions at an energy $\sqrt{S_{NN}} \approx$ 10 GeV is $L \approx 10^{27}$ cm^{-2} s^{-1}, and for proton–proton collisions at $\sqrt{S_{NN}} \approx 20$ GeV it will amount to $L \approx 10^{30}$ cm^{-2} s^{-1}. At present, four world's accelerator centers, namely CERN (SPS), BNL (RHIC), GSI (FAIR), and JINR (NICA), plan experimental programs in this energy range and complement each other, because any experimental results require confirmation.

The scientific goal of the project is to study the properties of heavily compressed baryon matter at the highest density attainable with accelerator technology (see figures 4.11 and 2.44) in an effort to discover and investigate new nuclear matter states and new phases of the quark–gluon plasma type.

In the NICA project, electron-nuclear collisions will be also investigated for the purpose of studying the electromagnetic form factor of nuclei and nucleons for high momenta. This line of research comprises studies of spatial charge distribution, magnetization of nucleons and nuclei, as well as parton distribution caused by valence quarks. Measurements of this kind will perhaps yield information about the color transparency of the baryon substance, which is important for understanding the properties of the quark–gluon plasma and the predictions of quantum chromodynamics at zero temperature.

Although the NICA energy range is much lower than that for RHIC and LHC, this project will permit realizing higher baryonic matter densities precisely where the phase transitions to quark–gluon plasma and chiral structures are expected to occur.

Clearly, the NICA project is capable of producing data on the role of non-equilibrium and dimensional effects in relativistic collisions, too.

A study will also be made of the mechanisms of multiparticle production, whereby the hadrons of a nucleus decay into final particles. At high energies (above 15 GeV), this effect is adequately described using quark-and-gluon models. At the same time, these transitions have not been adequately investigated at low energies and will supposedly call for invoking new degrees of freedom.

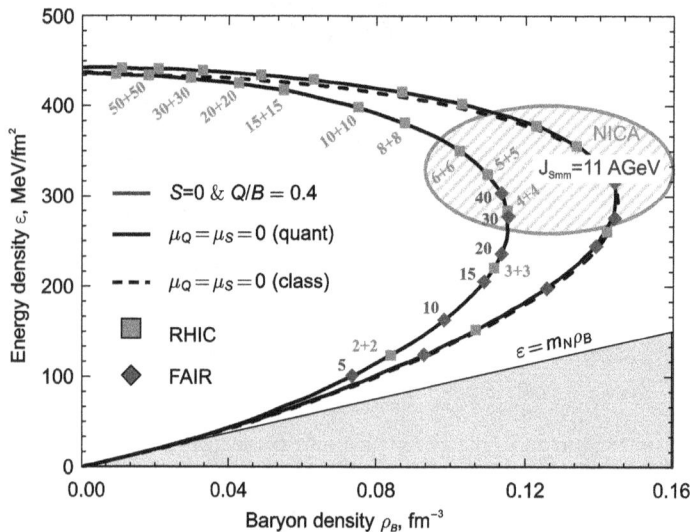

Figure 2.44. Freeze-out (ceasing of particle interactions in the system) estimated for different colliding energies [40]. The freeze-out baryon density is maximal at the colliding energy $\sqrt{S_{NN}} = 4 + 4$ eV. For the RHIC and FAIR, the energies are given in the laboratory frame, and in the center-of-mass system, respectively.

An important part of the NICA project is the physics of multiplicative effects, which is described by quantum chromodynamic techniques at a high density in the nonperturbative mode [40].

Cumulative processes, which are kinematically forbidden for free nucleons, may yield interesting information about collective properties of the medium and about short-range correlations of quarks and gluons in hot and cold quark–gluon plasmas.

The authors of the NICA project hope that they will be able to obtain new information about the properties of vacuum and hadrons. The matter is that the topological solutions of quantum chromodynamics, which arise from the non-Abelian nature of this theory, may give rise to topological fluctuations in the transcritical domain and, as a consequence, to spatio-temporal domains with broken P- or CP-invariance.

2.5 Technical applications

2.5.1 Magnetic confinement fusion

The history of controlled fusion is impressive and dramatic; it is full of hopes and disappointment, bright breakthroughs and failures. In his concluding speech to the 1961 Conference in Salzburg concerned with achievements in plasma theory, M N Rosenbluth said 'Let me close by saying that while it is unfortunately true that theorists have not told the experimentalists how to build a thermonuclear machine, it is also true that we have been looking hard for very many years for a fundamental reason why a plasma fusion reactor should be impossible and we have not found any such reason.' Next he added '... I believe the chances are very good that in twenty years or so mankind will have solved the problem of controlled fusion if only he has not lost in the meantime the far more difficult struggle against uncontrolled fusion.'

Now, 60 years after the commencement of controlled nuclear fusion research we may conclude [41] that the complexity of the problem was strongly underestimated in the initial stage of the work, especially so when it is considered that the final objective, namely the demonstration of electric power production by a thermonuclear power plant, is still several decades away.

For practical realization of controlled fusion with magnetic confinement, two methods of heat insulation of hot plasma are considered, i.e. magnetic and induction. In the former case, which has been given priority, the plasma with thermonuclear parameters—a temperature of $\approx 10^8$ K and a density of 10^{19}–10^{20} m^{-3}—is confined by magnetic fields of induction 4–8 T in a continuous or quasi-continuous mode. This line has received the name 'magnetic confinement'. Its leader is a toroidal tokamak facility, which was developed at the I V Kurchatov Institute of Atomic Energy in the 1950s–1960s and which underlies the International Thermonuclear Experimental Reactor (ITER) constructed jointly by seven countries in France. The project was formulated with the inclusion of a vast base of experimental data and simulation results obtained with model codes that took into account the main processes in the tokamak.

The reader may familiarize himself with the works on plasma heating and its magnetic confinement in review [41]. Here, we mention only several results.

The most impressive event consisted in the release of substantial thermonuclear power in deuterium–tritium plasma experiments on the TFTR (11 MW, 1994) and JET (16 MW, 1997) tokamaks (figure 2.45). The highest value of $Q = P_{fus}/P_{aux}$ achieved on the JET facility was about 0.65. These results were recorded in the modes with hot ions, $T_i \gg T_e$, which are not typical of the reactor. In the reactor-like H-mode on the JET facility with $T_i \approx T_e$, a fusion power $P_{fus} = 3$–5 MW was obtained in a long (≈ 5 s) pulse. Similar results were achieved on the JT-60U facility in deuterium discharges: the equivalent value of Q_{eqv} calculated for a DT plasma reached approximately 1.25 in a short pulse at $T_i \gg T_e$ and about 0.5 in the quasi-stationary mode.

Figure 2.46 shows the maximum thermonuclear power measured in DT discharges or the equivalent power calculated from the parameters of the DD plasma in various tokamaks, P_{fus}^{max} as a function of time in the interval from 1975 to 1995.

One can see that P_{fus}^{max} has increased by 10^8 times over a period of 20 years. This was achieved by constructing new, larger facilities and equipping them with higher-power additional heating. On obtaining the record-high power pulses on the JET and JT-60U facilities, no further increase occurred in P_{fus}^{max}. The new superconducting facilities constructed during the last decade, which are smaller in size than the

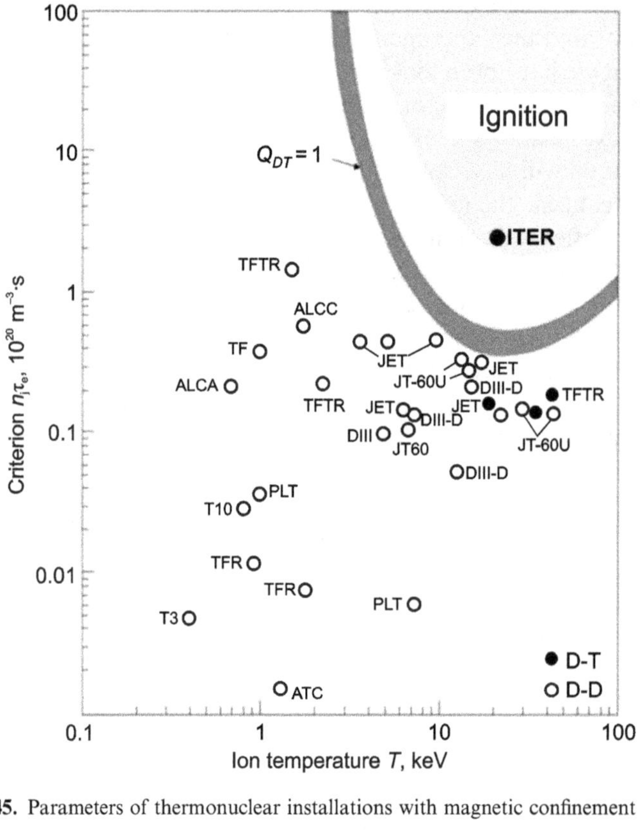

Figure 2.45. Parameters of thermonuclear installations with magnetic confinement of plasma.

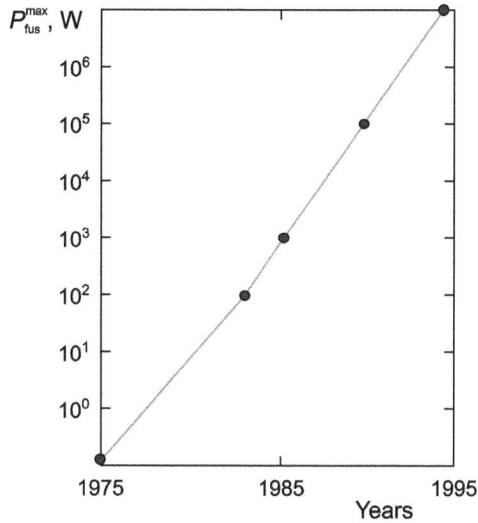

Figure 2.46. Growth dynamics of fusion power generated in different experimental facilities over a period of 20 years (1975–95).

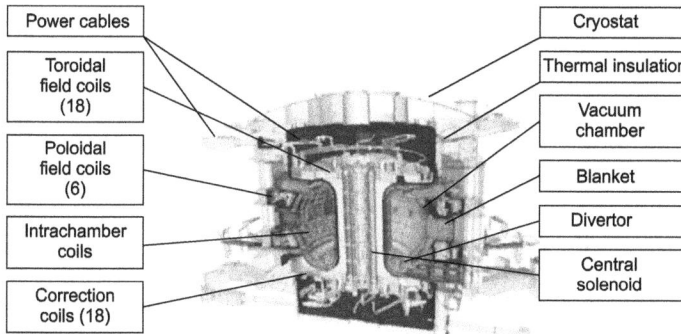

Figure 2.47. International thermonuclear experimental reactor (ITER). Reprinted from [18] by permission from Springer. Copyright 2011.

JET and JT-60U, are intended for the realization and investigation of stationary discharges rather than the attainment of high $P_{\text{fus}}^{\text{max}}$ values. The further increase in $P_{\text{fus}}^{\text{max}}$ (by a factor of 30–50 in comparison with the values attained on the JET and TFTR) should occur when the ITER reaches its design parameters, i.e. by 2027.

The ITER project (figure 2.47) is designed to produce a DT plasma with $P_{\text{fus}} = 400$–500 MW and $Q \geqslant 10$ in the induction mode with a pulse duration of about 500 s. The feasibility of achieving the 'controllable DT-plasma combustion', i.e. the modes with $Q > 30$, should also be investigated. It is supposed that the ITER will be the last physical facility, at which it will be possible to demonstrate the tokamak plasma confinement modes for an intense fusion reaction (thermonuclear plasma burning). It is expected that the emergence of a significant alpha-particle population in the tokamak plasma will give rise to the build-up of new instabilities, so-called Alfven modes, because of the nonequilibrium character of the ion distribution function.

It will be necessary to study the effect of these instabilities, in particular, on the plasma and alpha-particle confinement. An increased loss of the alpha-particles, which heat the plasma, will not permit lowering the power of additional heating sources in the stationary mode of reactor operation. The existing theories do not predict a catastrophic effect of the Alfven modes on the plasma stability and alpha-particle loss. However, an experimental confirmation will be required. Therefore, the main physical task of the ITER is to study the physics of thermonuclear plasma burning in a tokamak reactor.

Another ITER mission is to demonstrate the operating capacity of the crucial technologies of a tokamak thermonuclear reactor: control over the modes of operation and over additional heating and instability suppression, tritium handling, superconducting magnetic system operation.

Apart from tokamaks, other facilities with magnetic confinement are also investigated at the world's laboratories. Of most interest is a stellarator (figure 2.48), which offers several advantages over tokamaks. The main virtue of the stellarator is that the rotational transformation of the magnetic field lines is achieved by selecting magnetic field coils with a complex three-dimensional geometry. That is why there is no need to excite the current for confining the stellarator plasma and, hence, there are no magnetohydrodynamic instabilities arising from the current flow. The stellarator is free from the most hazardous tokamak instability—the disruption instability. The physical and technical characteristics of the stellarator are so far less developed than those of tokamaks, and the design of its magnetic system is highly complicated and is poorly suited for the accommodation of a blanket intended for absorbing a neutron flux. The stellarator divertor used for removing alpha particles and impurities from the plasma is radically different from a tokamak divertor and is technically more sophisticated. The accommodation of the thermal insulator intended to protect the cryogenic superconducting magnetic system from the neutron flux is also a major problem. As a consequence, in conceptual projects of a reactor stellarator the major chamber radius is long, of the order of 15 m, and the

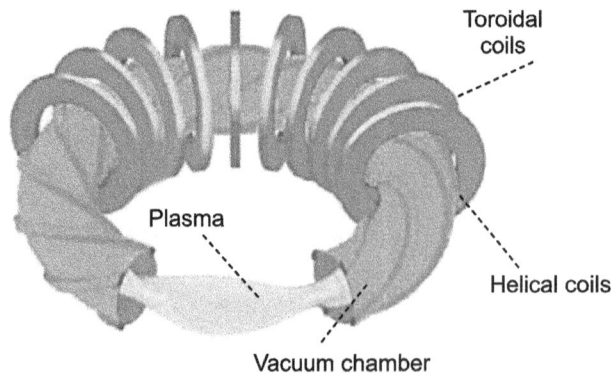

Figure 2.48. Schematic representation of a stellarator. Reprinted from [18] by permission from Springer. Copyright 2011.

output electric power amounts to several gigawatts, which is too high for an industrial power plant.

2.5.2 Laser inertial confinement fusion

Direct-drive laser fusion. Controlled thermonuclear fusion with inertial confinement relies on the feasibility of obtaining positive thermonuclear energy release in the form of microexplosions initiated by laser, x-ray, or heavy-ion radiation with an energy of several megajoules and a duration of about 1 ns. This line of research is the most significant pragmatic motivation for studies in high-energy-density physics [4, 7, 11]. To this end, the world's largest NIF laser facility (Livermore, USA) was put in operation with the pulse energy amounting to 1.8 MJ in March 2012; another megajoule laser facility, LMJ (Bordeaux, France) [6] is being constructed; different new schemes of high-current Z pinches are under investigation; planned for the future, in the framework of the FAIR project [37], is the commissioning of a new-generation relativistic heavy-ion accelerator. These facilities will be used (along with the thermonuclear program) for simulating the physical processes in thermonuclear weapons and studying nuclear physics.

The temperature dependence of the rate of thermonuclear reactions (figure 2.49) makes the deuterium–tritium fusion reaction most preferable; a temperature of 2–10 keV is needed for the reaction to be efficient. For a thermonuclear plasma expansion velocity of 10^8 cm s^{-1}, this leads to a characteristic time of the order of 10^{-9} s in the case of a spherical target of about 10^{-1} cm in diameter. The energy balance condition for this thermonuclear fusion (the Lawson criterion [4]) has the form

$$n\tau \approx \rho r/4c_s m_i \approx 2 \times 10^{14} \text{ s cm}^{-3},$$

which corresponds to $\rho r \approx 0.1\text{–}3$ g cm^{-2} and requires the thermonuclear fuel to be compressed to densities of hundreds of grams per cubic centimeter. At a temperature

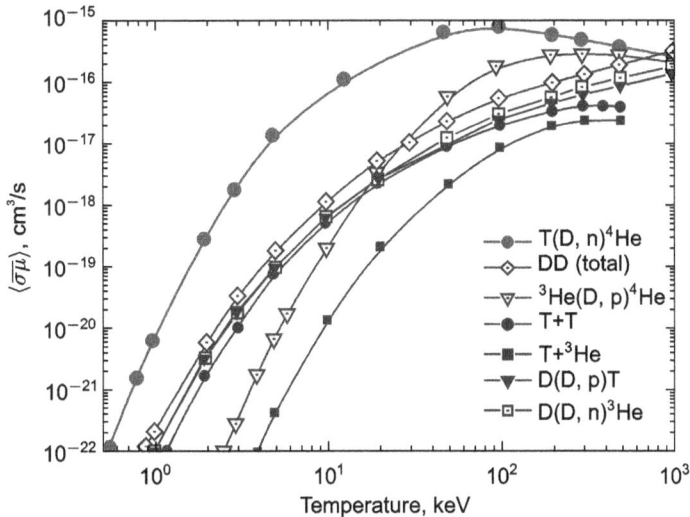

Figure 2.49. Dependences of the rate of thermonuclear reactions on temperature [4].

of 10 keV, this corresponds to tremendous pressures of 100–200 Gbar. Thus, to release a thermonuclear energy equivalent to 0.1 ton of trinitrotoluene, or 500 MJ requires three milligrams of the deuterium–tritium fuel, which corresponds to a microtarget $\approx$300 µm in diameter.

As a result of laser irradiation, the fuel density at the target center amounts to 90 g cm^{-3} for a compression energy of 30 kJ and a pressure of 13.5 Gbar, which may be achieved for an implosion rate of 1.4×10^7 cm s^{-1}. These parameters can vary significantly, depending on the specific design of the targets and the selected drivers.

A typical scheme for the operation of a simplified thermonuclear target is shown in figure 2.50 [4]. Under the action of a flux of directed energy from the 'driver', the plasma crown is heated and starts moving toward the incident radiation. The resulting recoil momentum then forms compression waves in the target, which are focused at the center of the target. At the moment of maximum compression, thermonuclear combustion begins at the center of the target and then propagates in the form of a wave from the center of the target to the periphery. Figure 2.51 [4] shows the distribution of the parameters in this target. This scheme of controlled thermonuclear fusion with inertial confinement is referred to as direct-drive laser fusion, because the target is compressed and heated under ablation plasma pressure ($p \approx 100$ Mbar, $T \approx 10^6$ K, figure 2.52) produced at its outer surface by focused laser radiation. To date, a large number of direct-drive thermonuclear target designs have been put forward (see figure 2.50), where the energy density is as high as 10^{16}–10^{17} J cm^{-3}, which is comparable with astrophysical conditions.

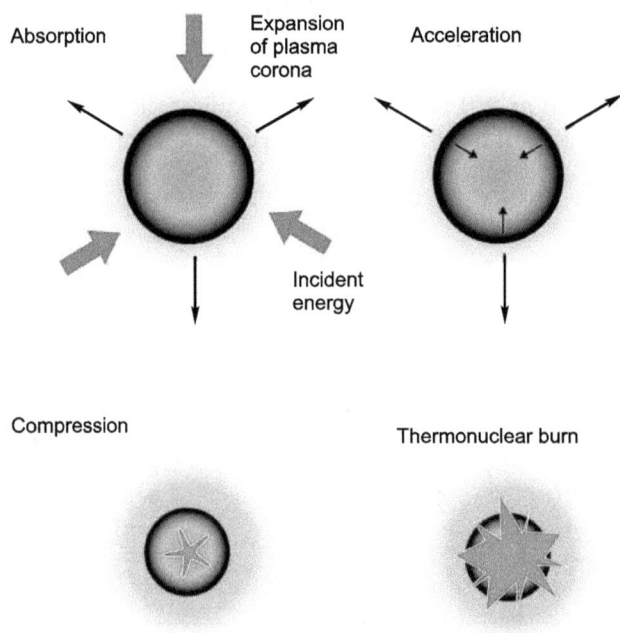

Figure 2.50. Schematic representation of thermonuclear target operation. Reprinted from [18] by permission from Springer. Copyright 2011.

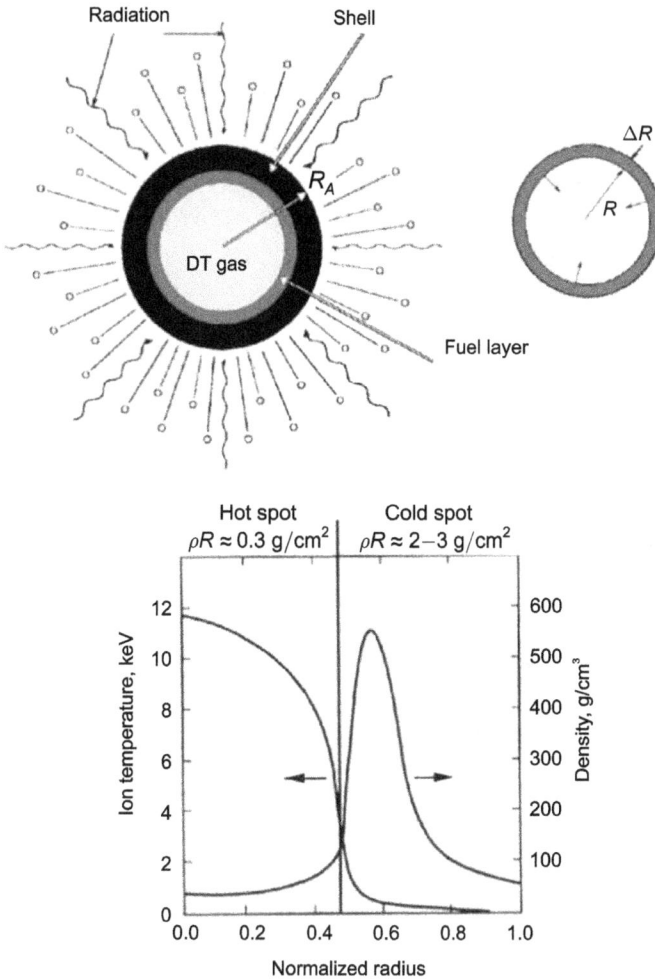

Figure 2.51. Variation of parameters in a thermonuclear target. Reprinted from [18] by permission from Springer. Copyright 2011.

An important problem of the practical implementation of pulsed controlled thermonuclear fusion, apart from the development of a high-power 'driver', is to ensure a high symmetry of irradiation and, accordingly, symmetry of the dynamic plasma compression. To reduce the distortion effect of the Rayleigh–Taylor instability in the course of laser-driven compression, a so-called indirect-drive scheme [4] has been developed (figure 2.53), where the spherical target (figure 2.53(b)) is compressed by the thermal soft x-rays from the side walls of a cylindrical capsule, heated by laser radiation.

In the scheme (figure 2.53) and photograph (figure 2.54) of the thermonuclear NIF target, a spherical DT target (figure 2.53(b)) is placed in a gold cylinder irradiated from two sides by 192 laser beams; upon incidence on the inner surface of the cylinder (see figure 2.53(a)) (at a power density of the order of 10^{15} W cm^{-2}), the beams evaporate it to produce high-intensity soft x-rays in the cylindrical cavity.

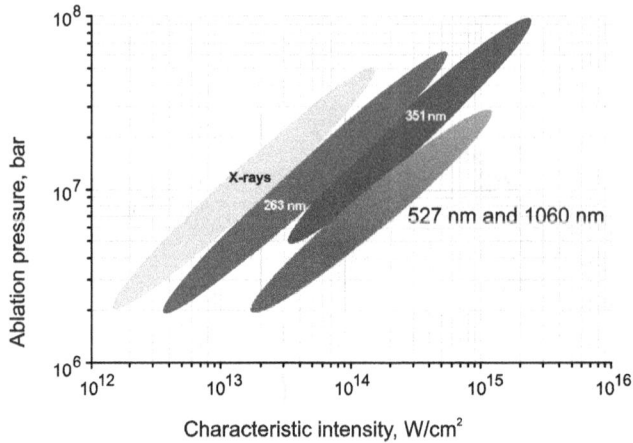

Figure 2.52. Dependence of ablative pressure on intensity and wavelength of electromagnetic radiation of the 'driver' [4]. Reprinted from [18] by permission from Springer. Copyright 2011.

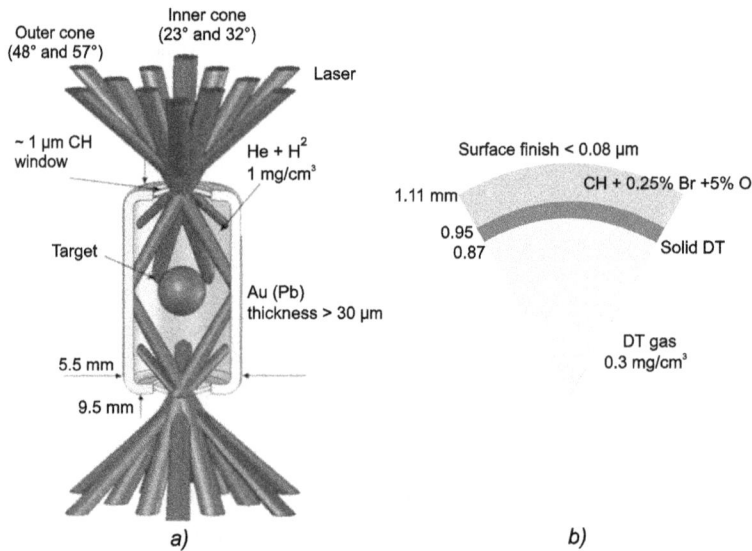

Figure 2.53. (a) Schematic representation of an indirect-drive thermonuclear target for the NIF laser system and (b) cross section of the thermonuclear sphere inside the target irradiated by x-rays. Reprinted from [18] by permission from Springer. Copyright 2011.

These x-rays in turn cause highly symmetric spherical compression of the thermo-nuclear target (figure 2.53(b)). To eliminate the distorting influence of the wall plasma on the symmetry of compression, the interior of the cylinder is filled with a low-density ($n \approx 10^{21}$ cm^{-3}) plasma obtained by ionization of airgels. Similar schemes of thermonuclear targets for heavy ions are presented in papers [7, 8].

The operation of these thermonuclear targets is associated with complex non-linear processes, i.e. beam filamentation and self-focusing, incoherent scattering, generation of plasma waves (stimulated Raman and Brillouin scattering),

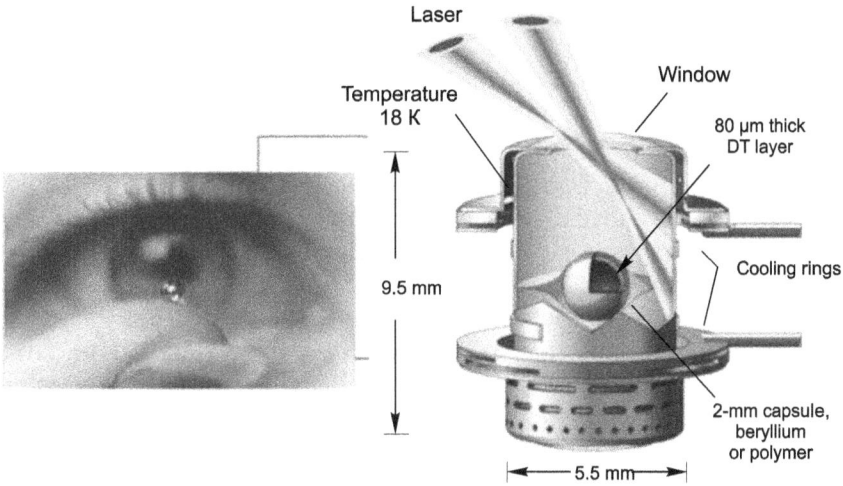

Figure 2.54. Photograph of an indirect-drive target and scheme of its irradiation by 192 laser beams of the NIF facility. Reprinted from [18] by permission from Springer. Copyright 2011.

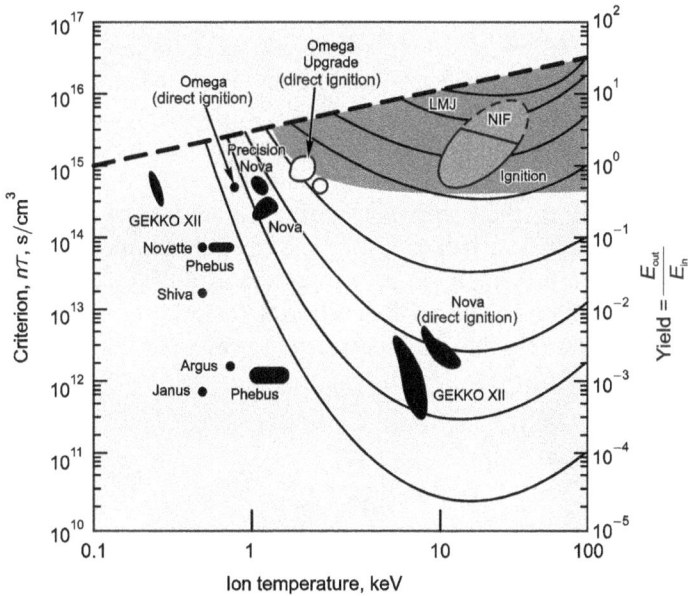

Figure 2.55. Capabilities of lasers for controlled thermonuclear fusion with inertial plasma confinement. Reprinted from [18] by permission from Springer. Copyright 2011.

nonthermal electrons and ions, two-plasmon decay, and many other phenomena, some of which have been adequately studied on smaller facilities and with the use of sophisticated computer codes.

The capabilities of existing laser systems for inertial thermonuclear fusion and of those under construction are illustrated by the data in table 2.4 and figure 2.55, which show that the commissioning of megajoule NIF and LMJ lasers will provide thermonuclear ignition conditions and energy-positive controlled nuclear fusion modes.

At the same time, the vigorous progress of laser technology constantly provides new technical solutions and physical ideas that strongly influence the development of the program of laser fusion. In particular, the realization of crystal and glass laser pumping by semiconductor laser diodes raises the total system efficiency to 30%–35%.

Fast ignition. The advent of ultra-power short-pulse lasers opens up new interesting possibilities for laser fusion, making it possible to separate the processes of adiabatic compression and heating of a thermonuclear plasma [9]. In this scheme, the thermonuclear target is compressed to high densities and heated by a nano-second laser pulse well below thermonuclear temperatures; then, the inner part of the compressed fuel is additionally heated by intense fluxes of megaelectronvolt electrons or ions arising from the femtosecond laser irradiation to initiate thermo-nuclear combustion of the entire target (figure 2.56).

In addition, at very high intensities of the laser beam, one can expect effects that contribute to a deeper penetration of light energy into the plasma of supercritical density. Due to the action of ponderomotive light-wave forces on the electrons, electrons and ions will be forced out of the laser beam zone, resulting (due to such self-focusing) in the formation of a low-density plasma channel through which the light energy will be delivered beyond the critical surface into the interior of the pre-compressed target and initiate its thermonuclear ignition (see figure 2.56). This scheme, which is referred to as 'fast ignition', exhibits a higher thermonuclear yield and, what is highly important, a higher immunity to instabilities and mixing, because the stages of compression and heating are separated in time in this case. The requirement to optimal target operation leads to the conclusion that the duration of the laser pulse that compresses the thermonuclear fuel to 200–300 g cm^{-3} should be equal to $\approx$10–20 ns.

The main problems of this scheme are associated with the efficient generation and transfer of ultrahigh-intensity energy fluxes into the interior of the compressed dense plasma. Estimates show [11] that it is necessary to heat a plasma volume of the order

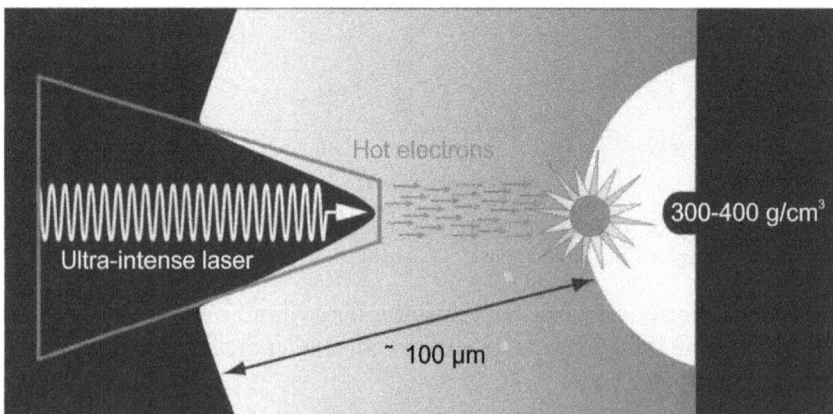

Figure 2.56. Scheme of fast ignition of a thermonuclear target. The electron current is more than 500 MA. Reprinted from [18] by permission from Springer. Copyright 2011.

of the electron or α-particle path ($\rho r \approx 0.5$ g cm^{-2}) to 10 keV, which is about 10 μm in a time of 10–20 ps at a plasma density of 300 g cm^{-3}. The corresponding energy is equal to ≈3 kJ for a power of 4×10^{14} W and an intensity of ≈10^{20} W cm^{-2}. In a different version of this scheme, for heating, use can be made of electrons or protons with an energy of 1–5 MeV, which are comparable to α-particles in range.

Modern short-pulse lasers provide the required intensities of ≈10^{15}–10^{20} W cm^{-2} and generate a wide spectrum of effects useful for fast ignition in 10^{21}–10^{26} cm^{-3} density plasmas—relativistic self-focusing and filamentation, quantum and sausage-type instabilities, as well as formation of vacuum channels and a set of new diverse particle acceleration mechanisms. Also possible is the generation of magnetic fields of superhigh intensity (≈10^9 G) and multimegaelectronvolt-energy ions.

An important role in the study of these effects is played by computer simulations, which predict, in particular, a highly efficient (up to 30%) energy transfer from laser radiation to megavolt-range electrons (figure 2.57). In this case, there emerge several new acceleration mechanisms—heating by oscillating ponderomotive forces, trans-verse-to-longitudinal electric field transformation in superdense plasma layers, acceleration of electrons in the betatron resonance in relativistic laser channels.

All this variety of phenomena is the subject of careful studies. In particular, about 20% of the energy of a short-pulse laser was transferred to the compressed thermonuclear target in experiments conducted at the Osaka University [42].

The authors of [43, 44] came up with the idea of igniting the compressed fuel by fast ions of a laser plasma rather than electrons.

2.5.3 Heavy-ion beam fusion

Initiation of controlled thermonuclear reactions by beams of heavy (with atomic number higher than 80) ions [7] has a number of potential advantages, such as high

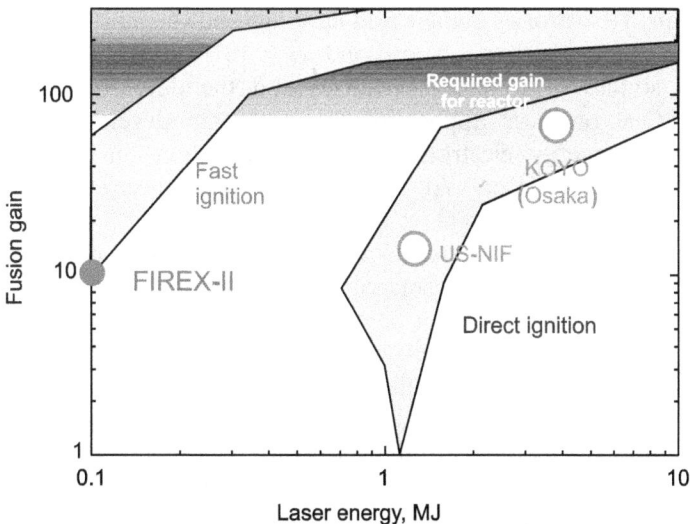

Figure 2.57. Thermonuclear gain coefficient versus laser energy. Reprinted from [18] by permission from Springer. Copyright 2011.

efficiency of accelerators, as well as reliability and maturity of ion accelerators developed for experiments in high-energy physics. Considered for this purpose is the multigigaelectronvolt level of ion kinetic energies for a total beam energy ranging into the megajoules; these beams should be compressed to 10 ns and focused onto millimeter-sized targets. For meeting these conditions, it is required to go beyond the limits defined by the volume charge, which calls for the suppression of diverse instabilities and the study of collective effects in high-current beams as well as of several other complex processes. Thus, the rise of the beam temperature may be caused by collective processes (up to beam 'crystallization'), the imperfection of magnetic field systems for the reflection from conducting surfaces, and inter-ion forces.

The propagation of the beam in the reactor chamber also calls for a careful analysis [11] of the dynamics of the background plasma and the reverse currents (flux and filamentation instabilities), the 'peeling off' of the beam by the background plasma, its photoionization, and so on.

Research into ion beam generation and high-energy-density physics involving ion beams is being vigorously pursued in several research centers: GSI, Germany; Berkeley, USA; ITEF, Russia [7, 8], where a wealth of interesting new data was obtained concerning the absorption of heavy ions by plasmas, shock-wave dynamics, isochoric target heating, plasma spectra, etc [8].

Of special interest is the project to use the GSI heavy-ion accelerator (SIS 100) in combination with the petawatt PHELIX laser (see figure 2.37).

Work on the use of heavy ions in high-energy-density physics is described in greater detail in recent review [8].

2.5.4 Laser-plasma acceleration of charged particles

Currently, many laboratories in the world have high-power short-pulse lasers (with a pulse duration τ_L shorter than a picosecond, $\tau_L < 10^{-12}$ s) (see table 2.4), which are employed in studies of different processes and the development of numerous applications. One of these applications involves the development of compact laser-plasma high-energy electron and ion accelerators on the basis of new-generation femtosecond lasers. At the same time, laser-accelerated electrons can then be used to make x-ray lasers on free electrons.

Discussed are several mechanisms of generation of fast electrons under the influence of a laser pulse on a plasma with a density much higher than the critical one [45]. If the laser pulse does not have a prepulse (high contrast ratio), then the laser radiation interacts with a solid-state density plasma having a sharp boundary. In this case, the mechanism of 'vacuum heating' is realized, as well as the so-called $v \times B$ mechanism (B is the amplitude of the magnetic field induction of the laser wave) caused by the longitudinal (along the direction of laser pulse propagation) ponderomotive force. The $v \times B$ mechanism becomes significant at relativistic intensities, when the electron oscillation energy is comparable to or higher than the electron rest energy, $mc^2 = 511$ keV.

There is also a mechanism for generating fast electrons in a plasma resonance at the critical plasma surface when the laser radiation has a projection of the electric field strength vector on the density gradient (usually under oblique incidence of laser radiation on the target) and the laser frequency coincides with the plasma frequency.

Unlike the ponderomotive $v \times B$ mechanism, the 'vacuum heating' and the resonance absorption mechanism emerge for nonrelativistic (substantially lower, $\alpha < 1$) intensities as well.

Another mechanism of fast electron generation in the subcritical plasma region in front of the target is considered [45], which operates due to the betatron resonance in the arising magnetic field. In this mode, electrons are accelerated by the transverse ultrarelativistic electric field of the laser wave in the direction of wave polarization, while the azimuthal magnetic field induced by the fast electron current produces the magnetic part of the Lorentz force. This force turns the electrons in such a way that they gradually reverse their direction of motion. For an exact betatron resonance, their reflection takes place at the moment the transverse electric field changes its direction, so that the electrons are permanently in the acceleration mode.

There also exist other electron acceleration mechanisms, which require specific experimental conditions, for example, acceleration in the wake wave. In the case of resonance absorption, the field near the critical plasma surface is substantially higher than the field of incident laser radiation.

We do not set ourselves the task of describing numerous laser acceleration mechanisms at length (see review [45]) but list different mechanisms of electron heating in table 2.5.

Among the actively developed applications of laser-plasma ion acceleration, mention should be made of fast ignition in inertial thermonuclear fusion

Table 2.5. Comparison of different mechanisms of electron heating in dense media [45].

Heating mechanism	When it applies
Stimulated inverse bremsstrahlung in the scattering of electrons from ions	Intensities below 10^{15} W cm^{-2}
Longitudinal ponderomotive electron acceleration in the skin layer	Relativistic intensities above 10^{19} W cm^{-2}
Vacuum heating	High contrast ratio, moderate intensities, short pulses
Resonance absorption of laser radiation	Low contrast ratio, long pulses
Electron acceleration by a laser wake wave	Gas targets, substantial subcritical plasma region, ultrashort pulses
Cyclotron mechanism	Presence of an external constant magnetic field
Betatron mechanism	Vortex electric field produced by a varying magnetic flux penetrating the orbit of electrons

(see subsection 2.5.2) and proton therapy of malignant tumors (figure 2.58) [42]. The use of protons in radiation therapy and oncology has a number of significant advantages in comparison with other types of radiation. First of all, this is due to the fact that the deceleration of protons of a given energy in a substance takes place in a localized region in the vicinity of the Bragg peak [7, 8], where the main fraction of the beam energy is released, which reduces the irradiation dose of healthy parts of the body.

It is noteworthy that the necessary condition for the successful use of accelerated ions for the above purposes is their high monoenergeticity. In particular, the energy spread in the proton beam intended for hadron therapy should not exceed 2%. To produce quasi-monoenergetic beams of accelerated protons, two-layer targets were proposed, the analysis of which in real three-dimensional geometry was performed as described in paper [46], where it was demonstrated that it is possible to obtain well-localized bunches of high-energy protons with an energy spread $\Delta E/E \approx 3\%$ (see figure 2.59 [46]).

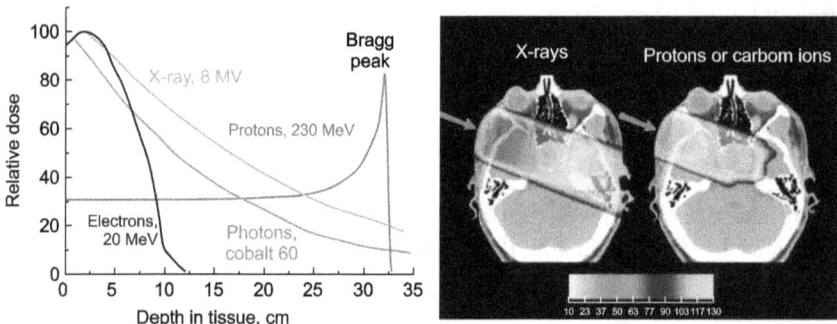

Figure 2.58. Proton and hadron therapy. Reprinted from [18] by permission from Springer. Copyright 2011.

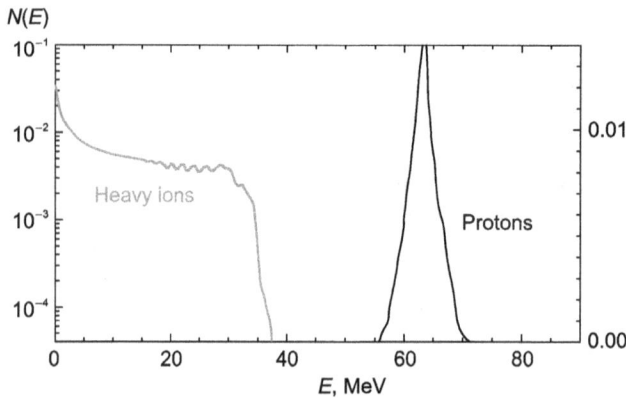

Figure 2.59. Spectrum of accelerated protons and heavy ions. Reprinted from [18] by permission from Springer. Copyright 2011.

2.5.5 Synchrotron sources, free-electron lasers, and terahertz generators

At present, free-electron lasers [47, 48] permit extremely intense radiation fluxes of femtosecond duration to be obtained in a broad (from the far-ultraviolet to soft x-ray) wavelength range (figure 2.60 [47, 48]).

These circumstances make free-electron lasers a promising tool for studying ultrafast chemical and biochemical processes with atomic-scale spatial resolution, for analyzing biological structures, for studying in real time the kinetics of physicochemical transformations, for investigating strongly nonideal and astrophysical plasmas, and for many other applications requiring atomic-scale spatial resolution and high temporal resolution.

Figure 2.61 shows a scheme of the FLASH free-electron laser at DESY, Germany [47, 48].

Today, this facility is the champion in radiation brightness, and the brightness of this device expected to be reached in 2013 will exceed that of the majority of facilities of this kind by six–nine orders of magnitude. In this case, the pulse duration will

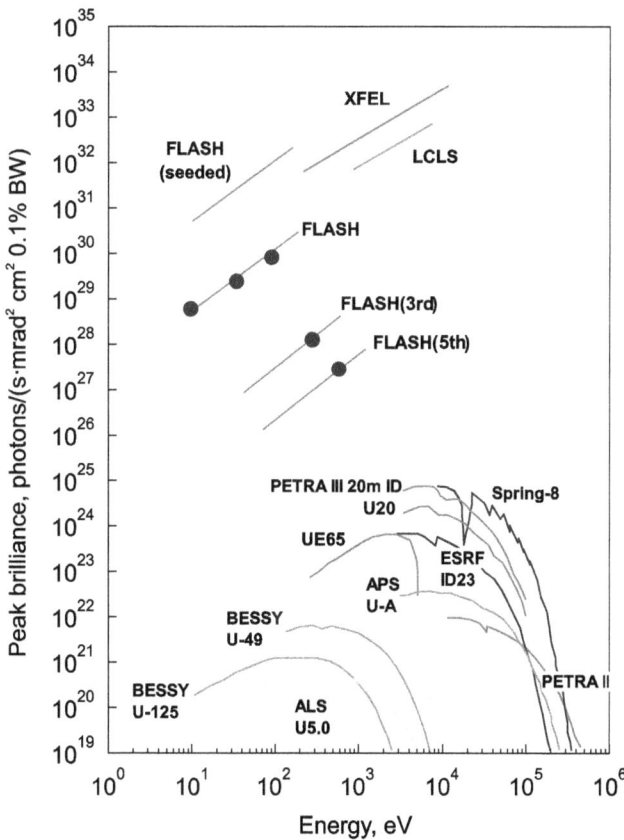

Figure 2.60. Peak power of experimental facilities FLASH and XFEL, Germany; LCLS, APS, USA; Spring-8, Japan; ESRF, France; and SLS, Switzerland. Reprinted from [18] by permission from Springer. Copyright 2011.

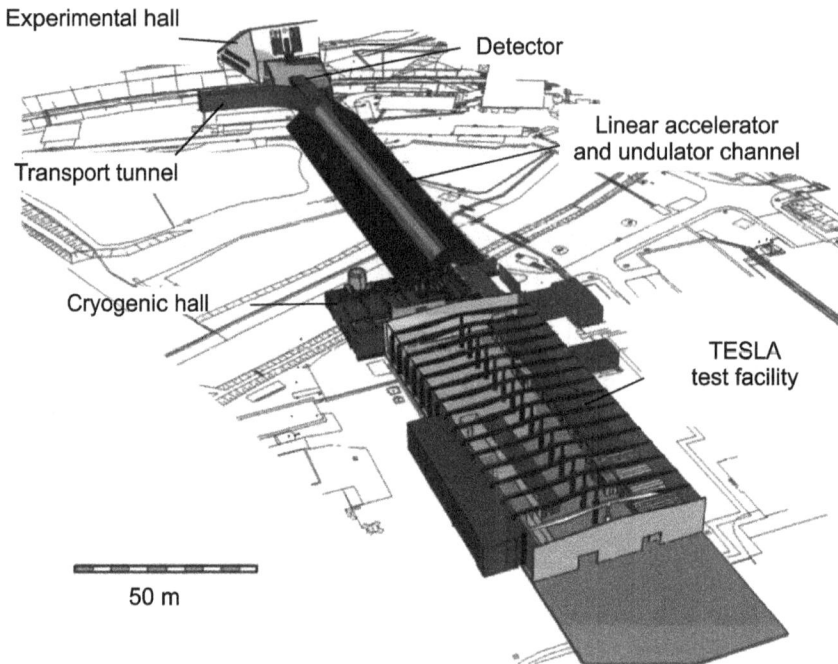

Figure 2.61. Scheme of the FLASH free-electron laser [47, 48]. Reprinted from [18] by permission from Springer. Copyright 2011.

decrease from 50 to 10 fs for a total pulse energy of 10–50 mJ and a wavelength of 6.5–60 nm. Recall that the light traverses a distance of only about 3 μm in 10 fs. The 260 m-long FLASH facility with a power of 5 GW consists of a superconducting linear electron accelerator with a current of 1–2 kA, which imparts to the electrons a kinetic energy of 440–700 MeV and an energy-spread of ≈0.1% This electron flux is applied to a 27 m-long undulator section, which generates x-ray electromagnetic radiation in the form of trains of 800 pulses with an energy up to 50 μJ each (150 μJ at maximum), with a repetition rate of 5–10 Hz, an average radiation power of 100 mW, and a peak brightness of 10^{29}–10^{30} photons/(s mrad mm^2 0.1%BW) [48].

With their unique radiation parameters, free-electron lasers permit a broad spectrum of investigations to be performed in different areas of science and technology that are quite frequently inaccessible by other means of generation and diagnostics.

Due to the short wavelength (comparable to the size of atoms), this radiation is suitable for studying the structure of individual atoms and molecules, whereas the short duration of radiation makes it possible to follow the kinetics and mechanism of chemical and biological reactions by exciting selectively the chosen degrees of freedom. This will enable, in particular, studies of three-dimensional biological structures as well as determination of the location and role of its specific constituent structural elements, which is of great importance in developing new-generation drugs and polymers, as well as in constructing complex spatial molecular structures.

In the more distant future, the application of this technique is expected to enable researchers to trace the variation in the electronic spectrum, magnetic properties, reactivity, and other physicochemical properties of clusters under continuous variation of the number of their constituent atoms from several atoms to solid-state values of the order of 10^{23} atoms.

References

[1] Al'tshuler L V, Trunin R F, Krupnikov K K and Panov N V 1996 Explosive laboratory devices for shock wave compression studies *Phys.-Usp.* **39** 539

[2] Al'tshuler L V, Trunin R F and Urlin V D *et al* 1999 Development of dynamic high-pressure techniques in Russia *Phys.-Usp.* **42** 261

[3] Avrorin E N, Simonenko V A and Shibarshov L I 2006 Physics research during nuclear explosions *Phys.-Usp.* **49** 432

[4] Avrorin E N, Vodolaga B K, Simonenko V A and Fortov V E I S 1993 Waves and extreme states of matter *Phys.-Usp.* **36** 337–64

[5] Azizov E A 2012 Tokamaks: from A.D. Sakharov to the present (The 60-year history of Tokamaks) *Phys.-Usp.* **55** 190–203

[6] Belyaev V S, Krainov V P, Lisitsa V S and Matafonov A P 2008 Generation of fast-charged particles and superstrong magnetic fields in the interaction of ultrashort high-intensity laser pulses with solid targets *Phys. Usp.* **51** 793

[7] Blaschke D *et al* (ed) 2009 *Searching for a QCD Mixed Phase at the Nuclotron-Based Ion Collider Facility (NICA white paper)* http://theor.jinr.ru/twiki/pub/NICA/WebHome/ Wh_Paper_dk6.pdf

[8] Boriskov G V, Bykov A I and Dolotenko M I *et al* 2011 Research in ultrahigh magnetic field physics *Phys.-Usp.* **54** 421–7

[9] Bychenkov V Y, Rozmus W and Maksimchuk A *et al* 2001 Fast ignitor concept with light ions *Plasma Phys. Rep.* **27** 1017–20

[10] Cavailler C 2005 Inertial fusion with the LMJ *Plasma Phys. Control. Fusion* **47** B389–403

[11] Dremin I M 2009 Physics at the large hadron collider *Phys.-Usp.* **52** 541–8

[12] *ELI: the Extreme Light Infrastructure European Project* http://extreme-light-infrastructure.eu/

[13] Emel'yanov V M, Timoshenko S L and Strikhanov M N 2011 *Vvedenie v Relyativistskuyu Yadernuyu Fiziku (Introduction into Relativistic Nuclear Physics)* (Moscow: Fizmatlit)

[14] Esirkepov T Z, Bulanov S V and Nishihara K *et al* 2002 Proposed double-layer target for the generation of high-quality laser-accelerated ion beams *Phys. Rev. Lett.* **89** 175003

[15] Fortov V, Iakubov I and Khrapak A 2006 *Physics of Strongly Coupled Plasma* (Oxford: Oxford University Press)

[16] Fortov V E (ed) 2000 *Entsiklopediya Nizkotemperaturnoi Plazmy (Encyclopedia of Low-Temperature Plasma)* (Moscow: Nauka)

[17] Fortov V E (ed) 2007 *Explosive-Driven Generators of Powerful Electrical Current Pulses* (Cambridge: Cambridge International Science)

[18] Fortov V E 2011 *Extreme States of Matter. Series: The Frontiers Collection* (Berlin: Springer)

[19] Fortov V E I S 2007 Waves and extreme states of matter *Phys.-Usp.* **50** 333

[20] Fortov V E, Hoffmann D H H and Sharkov B Y 2008 Intense ion beams for generating extreme states of matter *Phys.-Usp.* **51** 109

[21] Fortov V E, Sharkov B Y and Stöcker H 2012 European Facility for Antiproton and Ion Research (FAIR): the new international center for fundamental physics and its research program *Phys.-Usp.* **55** 582–602

[22] Garanin S G 2011 High-power lasers and their applications in high-energy-density physics studies *Phys.-Usp.* **54** 415–21

[23] Ginzburg V L O 1995 *Fizike i Astrofizike (About Physics and Astrophysics)* (Moscow: Byuro Kvantum)

[24] Grinevich B E, Demidov V A, Ivanovsky A V and Selemir V D 2011 Explosive magnetic generators and their application in scientific experiments *Phys.-Usp.* **54** 403–8

[25] Hammel B A National Ignition Campaign Team 2006 The nif ignition program: progress and planning *Plasma Phys. Control. Fusion* **48** B497–506

[26] Henderson D (ed) 2003 *Frontiers in High Energy Density Physics* (Washington: National Research Council, Nat. Acad. Press)

[27] HiPER http://hiper-laser.org/

[28] Hogan W J (ed) 1995 *Energy from Inertial Fusion* (Vienna: IAEA)

[29] Khazanov E A and Sergeev A M 2008 Petawatt laser based on optical parametric amplifiers: their state and prospects *Phys.-Usp.* **51** 969

[30] Kirzhnits D A 1972 Extremal States of matter (ultrahigh pressures and temperatures) *Sov. Phys.-Usp.* **14** 512–23

[31] Kirzhnits D A, Lozovik Y E and Shpatakovskaya G V 1975 Statistical model of matter *Sov. Phys.-Usp.* **18** 649–72

[32] Kodama R, Norreys P A and Mima K *et al* 2001 Fast heating of ultrahigh-density plasma as a step towards laser fusion ignition *Nature* **412** 798–802

[33] Korzhimanov A V, Gonoskov A A, Khazanov E A and Sergeev A M 2011 Horizons of petawatt laser technology *Phys.-Usp.* **54** 9–28

[34] Kozyreva A, Basko M and Rosmej F *et al* 2003 Dynamic confinement of targets heated quasi-isochorically with heavy ion beams *GSI-2003-2 Annual Report* 27

[35] Lebedev S V and Savvatimskii A I 1984 Metals during rapid heating by dense currents *Sov. Phys.-Usp.* **27** 749–71

[36] Maksimov E G, Magnitskaya M V and Fortov V E 2005 Non-simple behavior of simple metals at high pressures *Phys.-Usp.* **48** 761

[37] Mochalov M A, Ilkaev R I and Fortov V E *et al* 2010 Measurement of the compressibility of a deuterium plasma at a pressure of 1800 GPa *JETP Lett.* **92** 300–4

[38] Moses E I 2010 The national ignition facility and the national ignition campaign IEEE Trans. on Plasma Sci. 38 684–9 *36th IEEE Int. Conf. on Plasma Science (San Diego, CA) May 31–Jun 05, 2009)*

[39] New FLASH brochure

[40] Piskarskas A, Stabinis A and Yankauskas A 1986 Phase phenomena in parametric amplifiers and generators of ultrashort light pulses *Sov. Phys.-Usp.* **29** 869–79

[41] Roth M, Cowan T E and Key M H *et al* 2001 Fast ignition by intense laser-accelerated proton beams *Phys. Rev. Lett.* **86** 436–9

[42] Sharkov B Y (ed) 2005 *Yadernyi Sintez s Inertsionnym Uderzhaniem (Inertial Confinement Nuclear Fusion)* (Moscow: Fizmatlit)

[43] Shpatakovskaya G 2012 *Kvaziklassicheskii Metod v Zadachakh Kvantovoi Fiziki (Quasiclassical Method in Problems of Quantum Physics)* (Moscow: LAP LAMBERT Academic Publishing)

[44] *The European X-ray Laser Project XFEL* http://xfel.desy.de/

[45] Trunin R F 1994 Shock compressibility of condensed materials in strong shock waves generated by underground nuclear explosions *Phys.-Usp.* **37** 1123

[46] Zababakhin E I and Zababakhin I E 1988 *Yavleniya Neogranichennoy Kumulyatsii (The Phenomena of Unlimited Cumulating)* (Moscow: Nauka)

[47] Zasov A V and Postnov K A 2006 *Obshchaya Astrofizika (General Astrophysics)* (Fryazino: Vek-2)

[48] Zel'dovich Y B and Raizer Y P 2002 *Physics of Shock Waves and High-temperature Hydrodynamic Phenomena* (Mineola, NY: Dover)

IOP Publishing

Lectures on the Physics of Extreme States of Matter

Vladimir E Fortov

Chapter 3

Lecture 3: Interaction of laser radiation with matter

3.1 Physical effects arising under high-power laser irradiation

The action of high-power laser radiation leads to new and strongly nonlinear physical phenomena in relativistic plasmas with pressures of gigabars, with electric fields of teravolts per centimeter, and gigagauss magnetic fields [1] (see figure 2.26).

Apart from the well-known effects of self-focusing, stimulated scattering, and front steeping, there are emerging new effects: light filamentation, relativistic and ponderomotive effects in hydrodynamics, as well as fully developed generation of nonthermal gigavolt electrons and multimegavolt ions in laser plasmas, which result in nuclear reactions [1–3]. Here, we are dealing with extremely short—femtosecond—laser pulse durations, during which the electromagnetic wave executes only a few oscillations.

In analyzing the physical effects that arise when the irradiation power increases, we will move upward along the laser intensity curve (see figure 2.26). Beginning with $q > 10^{14}$ W cm^{-2} (for $\lambda = 1$ μm), the amplitude pressures of laser-driven shock waves pass into the megabar range [1, 4–6], in accordance with the scaling law [4–6]

$$p(\text{TPa}) = 0.87[q(\text{W cm}^{-2})]^{2/3}[\lambda(\mu\text{m})]^{-2/3}.$$

Beginning with $q > 3.4 \times 10^{18}$ W cm^{-2}, the electric field strength $E = \sqrt{4\pi q/c}$ in a laser wave is comparable with the electric field strength of the nucleus $E_a = e/a_{\text{B}}^2 \approx 5 \times 10^9$ W cm^{-1} in the first Bohr orbit of hydrogen. To ionize the energy level U_i [eV] requires an intensity [W cm^{-2}]

$$q = \frac{4 \times 10^9 U_i^4}{Z^2}$$

Under these conditions, laser radiation ionizes the medium, which turns into a heated plasma. Interesting experiments on the generation of high-power laser-driven

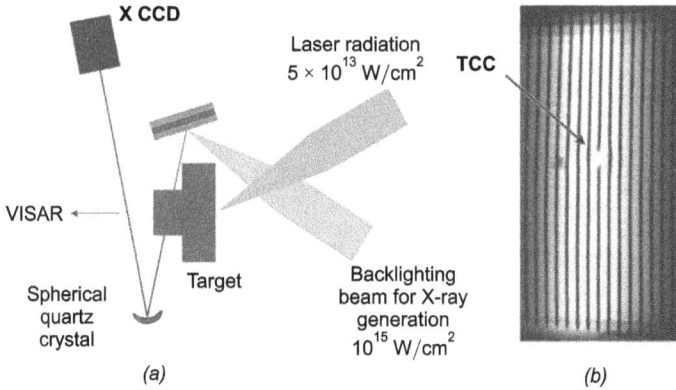

Figure 3.1. Experiments on (a) generation of laser-driven shock waves [7] involving the measurement (b) of shock compressed plasmas from the absorption of 5 keV x-ray radiation. Reprinted from [16] by permission from Springer. Copyright 2011.

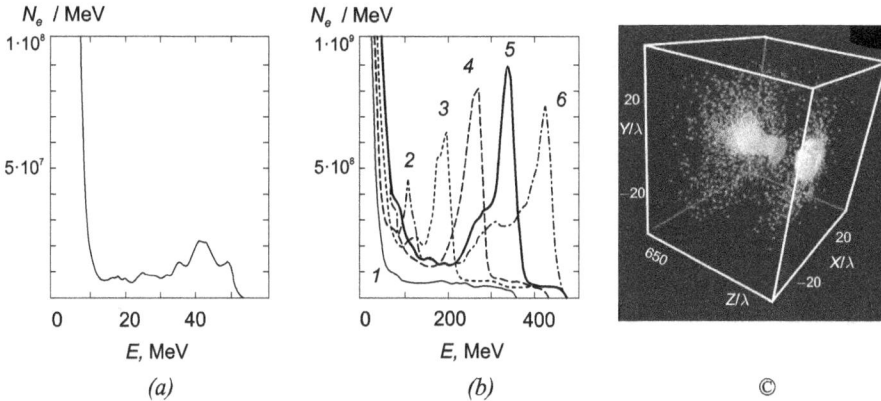

Figure 3.2. Energy spectrum of electrons accelerated by laser radiation [1]: (a)—20 mJ, 6.6 fs; (b)—12 J, 33 fs; temporal evolution of the spectrum: 1—$ct/\lambda = 350$, 2—$ct/\lambda = 450$, 3—$ct/\lambda = 550$, 4—$ct/\lambda = 650$, 5—$ct/\lambda = 750$, 6—$ct/\lambda = 850$; c—propagation of a laser pulse 12 J, 33 fs, $z/\lambda = 690$ through a plasma with a density of 10^{19} cm^{-3}. Three-dimensional picture of nonthermal electron energy distribution for $q \approx 10^{19}$ W cm^{-2}. Reprinted from [18] by permission from Springer. Copyright 2011.

shock waves [4, 7] (figure 3.1) and on the production of fast-charged particles in laser plasmas [1] (figure 3.2) are performed in this parameter domain.

It is noteworthy that the method of fast Thomson scattering can be used to construct the equation of state of a shock-compressed high-density plasma in experiments on the application of laser-driven shock waves. This substantially makes shock-wave experiments more informative in comparison with the ordinary technique of high-power shock waves [8, 9].

Beginning with roughly the same intensities $q > 10^{17}$ W cm^{-2}, an appreciable number of nonthermal electrons and ions of the multimegaelectronvolt range are generated in the absorption region [3]. Beginning with 10^{18} W cm^{-2}, the pondero-motive pressure of light is comparable with the hydrodynamic pressure of a plasma [1, 2, 6].

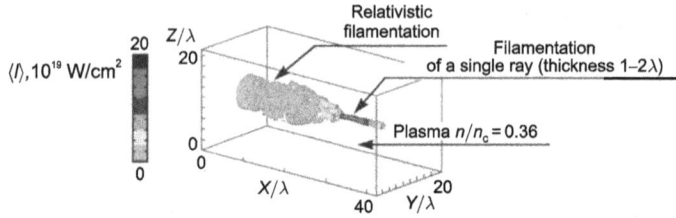

Figure 3.3. Self-focusing of a laser pulse. Reprinted from [18] by permission from Springer. Copyright 2011.

Relativistic effects become significant when the kinetic energy of an electron accelerated in the laser wave field is of the order of its rest energy $m_e c^2$, which leads to the condition

$$q_e \times \lambda^2 \approx 1.37 \times 10^{18} \text{ W } \mu\text{m}^2 \text{ cm}^{-2}.$$

For $\lambda = 1$ μm this gives 10^{18} W cm^{-2}.

Thus, microscopic quantities of matter with relativistic energies [1, 6] and with a relativistic electron mass of the order of 100 times the mass of the rest energy were obtained for the first time under terrestrial conditions.

The proton motion will become relativistic at intensities

$$q_p = \left(\frac{M_p}{m_e}\right)^2 q_e \approx 5 \times 10^{24} \text{ W cm}^{-2},$$

which, hopefully, will soon be realized in experiments.

The transition to relativistic intensities of laser radiation has already yielded several interesting physical results [10]. They include generation of x-rays, gamma-rays and betatron radiation, relativistic self-focusing in plasmas (figure 3.3 [1]) [11] and the atmosphere, generation of high-order harmonics, acceleration of electrons [1], protons and ions [3], generation of neutrons and positrons, generation of electron vortices and solitons [1], generation of ultramegagauss magnetic fields, as well as manifestations of quantum electrodynamics [6].

Under high-intensity irradiation, there occurs relativistic plasma 'transparentiza-tion' [1, 2], which is related to the relativistic growth of the electron mass and the corresponding decrease in the critical plasma frequency, $\omega_p = \sqrt{4\pi e^2 n_e / \gamma m}$ (where γ is the relativistic Lorentz factor), the plasma density modification by ponderomotive forces, as well as the frequency transformation of a laser pulse itself.

Along with the effect of relativistic plasma 'transparentization', of considerable interest is the effect of relativistic self-focusing of laser radiation caused by plasma permittivity variation due to the relativistic growth of the electron mass in the transverse direction relative to the beam propagation direction and the spatial plasma density redistribution under the action of ponderomotive forces. The critical power (in GW) for self-focusing has the form [11]:

$$W_{\text{cr}} = \frac{m_e c^5 \omega^2}{e^2 \omega_{\text{pe}}^2} \approx 17 \left(\frac{\omega}{\omega_{\text{pe}}}\right)^2.$$

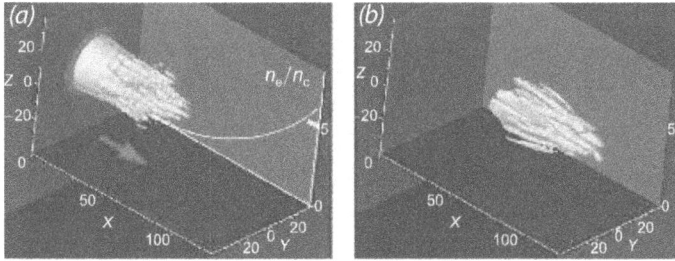

Figure 3.4. Multiple filamentation of a wide laser beam of petawatt power. Reprinted from [18] by permission from Springer. Copyright 2011.

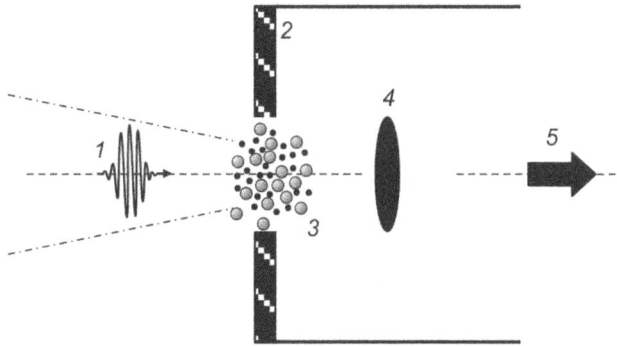

Figure 3.5. Schematic representation of the generator in question [12]: 1—laser pulse; 2—target; 3—plasma; 4—virtual cathode; 5—generated radiation.

Particle-in-cell (PIC) simulations [1] of laser beam filamentation are presented in figure 3.4 [1]. Calculations show that the initial pulse decays into several channels, which propagate quasi-independently due to their screening by plasma.

This multiray structure of the laser pulse arises due to the filamentation instability and the buildup of transverse beam nonuniformity, which resembles the self-focusing phenomenon. The self-focusing and filamentation of laser radiation in a plasma result in the formation of electron vortices, high-intensity compensative currents, and the generation of strong magnetic fields. The problem of mega- and gigagauss magnetic fields in laser plasmas is considered in detail in section 3.4.

In [12], a study was made of the feasibility of generating terahertz radiation (≈ 0.6 THz) under irradiation of targets by ≈ 0.1 ns laser pulses of high intensity $I \approx 10^{18}$–10^{19} W cm^{-2}. This investigation was performed by numerical simulations using a relativistic electromagnetic PIC code. The interaction of such a pulse with the target results in the generation of a plasma at the target (figure 3.5). The electrons escaping from the plasma form a virtual cathode, the oscillations of which are determined not only by their own field but also by the field of plasma ions. The generation proceeds in the terahertz frequency range, the generation efficiency being three times higher than that in the absence of ions, i.e. than with the traditional reditron generation mechanism, and amounts to about 10% of the electron beam power.

The state of the art and prospects of laser generation of terahertz megavolt pulses were analyzed in [13].

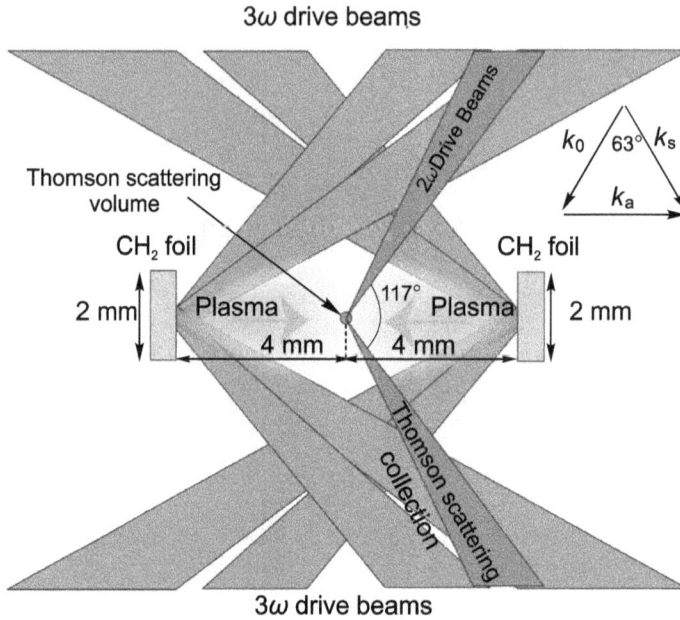

Figure 3.6. Experiment on laser generation of collisionless shock waves [14]. Each foil is irradiated by ten laser beams with a wavelength of ≈ 351 nm (3ω), a pulse duration of 1 ns, and a focal spot size of ≈ 250 μm.

Experiments [14] on the frontal collision of laser plasma flows to generate collisionless shock waves are shown in figure 3.6. CH_2 foils were irradiated by laser fluxes of intensity $\approx 10^{16}$ W cm^{-2} on the OMEGA laser facility. Using Thomson scattering techniques, it was shown that the ablated plasma had an electron temperature of ≈ 110 eV and a density of $\approx 10^{18}$ cm^{-3} for an expansion velocity of ≈ 2000 km s^{-1}. The frontal collision of such flows results in the generation of collisionless shock waves, which frequently occur in astrophysical plasmas.

The development of Kelvin–Helmholtz turbulence associated with collisionless shock waves was described in [15], where optical interferometric and proton radiographic techniques were used to study the dynamics of shock waves, contact discontinuities, and flows as a whole.

Interesting effects of the relativistic nature [1, 6] are related to strongly nonlinear plasma waves, which form vacuum channels and 'bubbles' (figure 3.7 [1]) in the plasma, produce plasma lenses for charged particles, and give rise to intense electromagnetic radiation in the frequency range from terahertz to x-rays, as well as excite collisionless shock waves.

The propagation of two collinear laser beams of relativistic intensity through a plasma leads to the generation of high-power electromagnetic wake waves enabling electrons to be accelerated (figure 3.2) with acceleration rates up to 100 MV cm^{-1}, which is thousands of times greater than ordinary acceleration gradients of about 5 kV cm^{-1} (see subsection 2.5.4). Gigantic longitudinal electric fields are generated in this case: for an intensity $q \approx 10^{18}$ W cm^{-2} the electric intensity is equal to about 2 TV m^{-1}, and for $q \approx 10^{23}$ W cm^{-2} it amounts to about 0.1 PV m^{-1}.

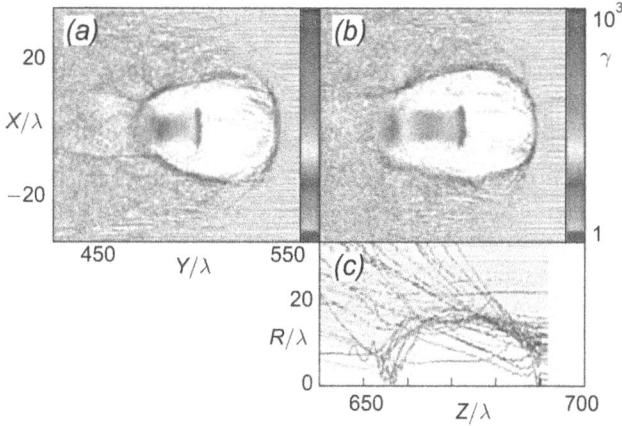

Figure 3.7. Soliton 'bubbles' in a plasma under irradiation by a 33 fs laser pulse with an energy of 12 J: (a)—$ct/\lambda = 500$; (b)—$ct/\lambda = 700$; (c)—electron trajectories in the frame of reference co-moving with the laser pulse [1]. Reprinted from [18] by permission from Springer. Copyright 2011.

These acceleration rates imply that the laser version of the ordinary 50 GeV SLAC accelerator would be only 100 μm long. Successful experiments have already been performed with laser-driven acceleration of electrons to energies of 1 GeV at laser radiation intensities of 10^{18}–10^{19} W cm^{-2}. There are reasons to believe that the advent of multipetawatt and exawatt lasers in the future would lead to the implementation of acceleration rates of the order of a teravolt per centimeter for a total kinetic energy ranging into the gigavolts (for more details, see subsection 2.5.4).

In the interaction of laser radiation of moderate intensity $q \approx 5 \times 10^{17}$ W cm^{-2} with frozen nanotargets, as described in [3], it was possible to obtain a flux of protons with energies of 5.5–7.5 MeV.

Among other interesting manifestations of nonlinearities in laser plasma, mention should be made of nonlinear steepening of the optical front (similar to the formation of a shock wave in hydrodynamics), plasma jet formation, and high-order harmonic generation, which is of practical interest for lithography, holography, medicine, etc.

Under the action of a circularly polarized electric field, the plasma electrons are set in circular motion to generate synchrotron radiation, relativistic effects being significant in its description at high intensities. For $\lambda = 1$ μm, this radiation-dominant mode is realized, beginning at intensities $q \approx 3 \times 10^{22}$ W cm^{-2}, when a substantial fraction of laser energy is radiated in the form of hard x-rays.

Quantum optical effects come to the fore at $q \approx 1.4 \times 10^{26}$ W cm^{-2}, the kinetic electron energy being equal to $\approx$50 TeV in this case. For $q \approx 10^{21}$ W cm^{-2}, the light pressure is equal to about 300 Gbar, which is close to the pressure in the center of the Sun and is much higher than the pressure in the near-source zone of a nuclear explosion [16–18].

High-intensity lasers make it possible to achieve superhigh acceleration $a_e = a_0 \omega c \approx 10^{30} g$ (for a dimensionless radiation amplitude $a_0 = eE/(m_e \omega c) \approx 10^5$), which is close to the accelerations in the vicinity of the Schwarzschild radius of a black hole [19] (see section 6.2). This permits the conditions in the neighborhood of black holes and wormholes to be modeled, and therefore the predictions of general relativity theory to be verified. Thus, for $q \approx 10^{26}$ W cm^{-2}, the electron acceleration amounts to $a = 10^{27}g$, which is close to the conditions of the black-hole event horizon [2, 19–21]. If such giant accelerations are realized, an opportunity is expected to open up to study the specific electromagnetic Unruh radiation, which is similar to the Hawking radiation caused by gravitational effects. In this case, additional electromagnetic radiation with an effective temperature $kT = ha/c$ should appear (in comparison with calculations using Maxwell's equations). The ratio of the power of this radiation to the synchrotron radiation power is equal to 10^{-6} at an intensity $q \approx 10^{18}$ W cm^{-2} and increases proportionally with $\sqrt{q}$, which raises expectations that it would be possible to observe it at high intensities of laser radiation.

At higher laser radiation intensities $q \approx 3 \times 10^{29}$ W cm^{-2}, there is a good chance to verify the predictions of modern quantum gravity theories [20, 22] about the change in space–time dimensionality at short distances. According to paper [22], this distance is $r_n \approx 10^{32/n-17}$ cm, where n is the dimension of space greater than four. In this case, the electron wave function will obey a different law of gravitation for $n < 3$ at distances of 10^{-6} cm.

The effects of quantum electrodynamics, polarization, vacuum breakdown, electron–positron pair production, and then quark–gluon plasma generation become significant at ultrahigh intensities of optical radiation, $q > 3 \times 10^{29}$ W cm^{-2}.

The problem of spontaneous electron–positron pair production in vacuum affects many interesting situations, such as the collisions of heavy nuclei (with $Z_1 + Z_2 > 135$), evaporation of black holes, particle production in the Universe, etc. The characteristic electric field scale [2] for the manifestation of the breakdown effect in quantum electrodynamics is the Schwinger intensity

$$E_S = \frac{m^2 c^3}{e\hbar} \approx 10^{16} \text{ W cm}^{-1},$$

which is sufficient to accelerate an electron to relativistic velocities at the Compton wavelength $\lambda_C = 2\pi\hbar/(mc)$ and corresponds to ultrahigh intensities of laser radiation

$$q_{QED} = q_e \frac{\lambda^2}{\lambda_C^2} \approx 8.1 \times 10^{30} \text{ W cm}^{-2}.$$

For these ultraextreme conditions to be realized, the focusing of $\lambda \approx 1$ μm laser radiation in a 1 mm^3 volume requires a release of ≈ 1 MJ energy. This intensity threshold for electron–positron pair production is substantially lowered (to 10^{22} W cm^{-2}) in the case of scattering by nuclei.

Further advancement along the laser intensity scale (figure 2.26) is hard to predict, for it is limited by our knowledge of the structure of matter in the immediate spatiotemporal neighborhood of the Big Bang at ultrahigh energy densities.

3.2 Laser-driven shock waves

Immediately after the advent of lasers, their unique properties were successfully used to generate high-power shock waves and to obtain extreme states of matter [4].

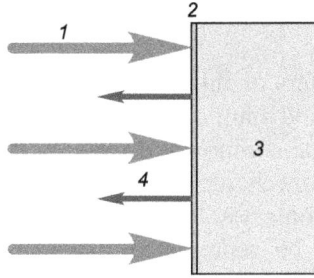

Figure 3.8. Scheme of the first experiments with the laser irradiation of a target.

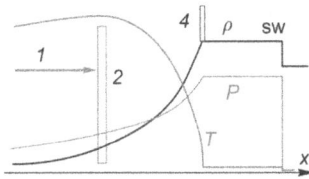

Figure 3.9. Profiles of density ρ, pressure p, and temperature T in the case of irradiation by a long pulse of sufficiently high power.

The scheme of the first experiments is shown in figure 3.8. Prolonged exposure of the target to a laser beam 1 produces a cloud of vaporized matter above its surface. The vapor expands along the arrows 4 towards the laser beam. If the vapor is transparent for radiation, then absorption takes place in the surface layer 2. This layer separates the target and the vapor. The vapor pressure 4 sets in motion the target material 3.

At moderate laser intensities, this is the vapor of neutral molecules or atoms, which is usually transparent to radiation [23], and the energy is absorbed in the surface layer of the target made of a condensed substance.

At high intensities I, the substance removed from the condensed target surface ionizes to form a plasma cloud (or a plume), which is referred to as a laser plasma corona. The issues of the corona dynamics have been studied in detail for many years in connection with the problem of laser inertial thermonuclear fusion, see, for example, papers [4, 24, 25]. Plasma is usually nontransparent to radiation, i.e. absorption takes place within the corona (with its density decreasing away from the target) in the vicinity of the critical surface, the zone where the electron density decreases to a critical value at which the plasma frequency decreases to the laser frequency. This density is about 10–100 times (depending on the hardness of a laser photon) lower than the solid-state density, with which the plasma is delivered to the corona.

The heat absorbed on the critical surface 2 (figure 3.9) is transported by the electron thermal conductivity mechanism to the target surface. 1 is the laser beam reaching the surface of critical density. The ablation front 4 forms the target boundary. This is the leading front of the thermal wave, beyond which the temperature rises sharply. The pressure of the corona enables the quasi-stationary propagation of the shock wave (SW) in the homogeneous substance of the target. The solid-density substance begins to intensively heat up at this front, its density lowers, and a new portion of the substance is delivered to the flow from the target to the corona.

By recording the parameters of this wave in experiments, it is possible to gain information about the thermodynamics of the shock-compressed substance.

Recording the shape of the compression pulse as it exits the free surface yields ample information about the mechanical properties of the medium at high pressures, temperatures, and deformation rates.

The resultant flow may be represented qualitatively as consisting of three portions: (1) a stationary shock wave followed by (2) a Chapman–Jouguet deflagration wave, where the light energy is absorbed, and (3) a simple centered rarefaction wave (adiabatic or isothermal) closes the flow.

In this case, the highest attainable pressure is

$$p_{\max} = I^{2/3} \rho_c^{1/3},$$

where I is the laser radiation intensity of frequency ω, and ρ_c is the critical plasma density ($\omega_0 = \omega_p = \sqrt{4\pi e^2 n_e / m_e}$).

An analysis of this relation suggests that the pressures attainable under laser irradiation depend only slightly on the chemical composition of the target. Herein lies a significant difference between the laser methods of pressure generation and the classical methods of dynamic physics that use the impact of metallic plates or the detonation products of condensed explosives. It is characteristic that high-frequency radiation has significant advantages in terms of obtaining maximum plasma pressures. However, the main advantage of short-wavelength radiation consists in a lowering of the effect of nonthermal electrons with increasing laser frequency.

One should bear in mind that the above qualitative estimate was obtained for the interval of laser pulse parameters in which the above-mentioned simplified inter-action model is purely schematic [4]. The point is that the absorption of light is essentially nonlinear at radiation intensities of 10^{13}–10^{17} W cm^{-2}. In this case, a substantial fraction of light is reflected from the plasma, and the reflection coefficient also depends on the radiation intensity. As a result, the formula for the absorption coefficient, used to estimate the screening time turns out to be oversimplified. At high laser intensities there arise a number of complications, which may be fully taken into account only in the framework of complex numerical simulations. The results of these simulations are approximated by the relation:

$$p_{\max} = \alpha I^\alpha \lambda_0^{-\beta},$$

where $\alpha \approx 0.3$–0.7 and $\beta \approx 0.3$–2.0, depending on the model and the radiation intensity.

The requirements to laser radiation and target size to produce plane and stationary shock discontinuities propagating through a relatively cold substance in question were formulated in [4]. This will permit using the dynamic diagnostic method based on the application of the conservation laws to the flow of shock-compressed plasmas. The limitations emerging in this case will define the highest pressure level attainable with the aid of modern lasers. Taking into account shock wave attenuation and curvature due to the rear and side unloading waves as well as considering the nonhydrodynamic target heating by nonthermal electrons, the authors of paper [4] obtained that the laser energy E_{las} at a wavelength λ_0 required for generating pressure p may be represented as:

$$E_{las} \sim p^6 \lambda_0^{11}.$$

The constructive way of advancing further along the pressure scale of laser-driven shock waves involves the use of shorter wavelengths and layered targets to suppress the role of electron heating. Therefore, in experiments with laser-driven shock waves, experimenters aspire to use the high-order harmonics of the fundamental radiation or laser-to-soft x-ray radiation conversion in 'holraum' schemes (see subsection 2.5.2).

Laser methods can be useful not only in the study of a highly compressed substance with a density exceeding the solid state. Using an expansion of a material heated by a laser-driven shock wave or nonthermal electrons, a wide range of states can be obtained in an isentropic unloading wave including the region of the Boltzmann strongly nonideal plasma, the vicinity of the high-temperature boiling curve, and the metal–dielectric transition region.

Even a brief enumeration of possible and already carried out experiments at very high local energy densities attainable with modern lasers shows that this technique has enormous advantages over other methods of obtaining high pressures and allows new physical information to be acquired about extreme states of matter. Naturally, such measurements are at the forefront of modern fast-response recording instruments and interpretation of these experiments calls for new physical models and complex numerical simulations.

We will consider below some experiments on laser generation of shock waves in solids and discuss the effects associated with this method of shock production.

The first experiments on the excitation of shock waves in solid hydrogen and plexiglass were carried out with a modest-power neodymium laser for an energy $E \approx 12$ J and a pulse duration $\tau = 5$ ns. Owing to the small (about 40 μm) size of the focal spot, the shock waves with a peak pressure of about 2 Mbar rapidly became spherical and decayed. For a radiation intensity $I = 3.5 \times 10^{14}$ W cm^{-2}, a shock pressure of ≈ 1.7 Mbar was obtained in polyethylene [26], and the amplitude pressures obtained in hydrogen and plexiglass for $I = 2 \times 10^{14}$ W cm^{-2} were equal to about 2 and 4 Mbar, respectively [27]. Measurements of the energy of plasma corona ions and the target recoil momentum (integral methods) made it possible to estimate the pressure in an aluminum target at $I \approx 10^{14}$ W cm^{-2}.

A higher-power Nd:glass Janus laser system with a pulse energy of up to 100 J and a pulse duration of 300 ps was used to generate plane shock waves [28, 29]. Radiation intensities I of up to 3×10^{14} W cm^{-2} were produced in focal spots 300–700 μm in diameter.

From the shock transit time through a stepped aluminum sample, it was possible to estimate the shock discontinuity front velocity of 13 km s^{-1} (corresponding to a pressure of about 2 Mbar) and measure the velocity of plasma corona expansion. Recording the temporal buildup of radiation intensity as the shock reached the free surface ($\Delta t \leqslant 50$ ps) permitted the thickness of the shock discontinuity to be estimated at less than 0.7 µm. The shock pressures were increased by an order of magnitude. The authors of paper [29] used a small diameter target to lower, in the authors' opinion, the effect of surface currents, and an agreement was obtained between theory [30] and experiment [29].

Plasma pressures $p \approx 35$ Mbar were obtained upon irradiation of a target consisting of a 22 µm thick aluminum layer and a 32 µm thick gold layer by ten overlapping beams of the Shiva neodymium laser facility with $\lambda_0 = 1.05$ µm [31]. The peak intensity, I, was equal to 2.9×10^{15} W cm^{-2} for a pulse duration of 625 ps. The velocity of the shock wave in gold was measured at 17.3 ± 0.3 km s^{-1} and was consistent with a two-dimensional hydrodynamic simulation [30], in which the fraction of the absorbed energy was equal to 30% and the laser beam convergence was taken into account. Measurements of the x-ray emission spectrum in this experiment made it possible to estimate the heating of the rear target side by nonthermal electrons. It turned out to be under 500 °C, while the temperature of the shock-compressed plasma was about 5 eV.

A version of the 'reflection' method was implemented in experiments [32] with a layered target, when a shock wave with an amplitude $p \approx 3$ Mbar underwent transition from aluminum to gold ($p \approx 6$ Mbar) under low-attenuation conditions. Like in [28], the laser radiation ($E \approx 20\text{--}30$ J, $\tau = 300$ ps) was nonuniformly distributed over the focal spot, its shape varying from circular (with a diameter of 100 µm) to elliptical (with axes of 200 and 500 µm). In the authors' opinion, the latter circumstance was the main source of errors in these experiments ($\delta D = 15\%$, $\delta p = 30\%$).

Systematic investigations of the shock compressibility of aluminum and copper were carried out using a comparative method on the Janus laser facility in the intensity range $I \approx 5 \times 10^{13}\text{--}4 \times 10^{14}$ W cm^{-2} ($E \approx 30$ J, $\tau = 300$ ps). A thin layer of gold in the target was employed to absorb nonthermal electrons and increase the duration of the shock wave (somewhat lowering, however, the peak pressure). The resultant data refer to a 2–6 Mbar pressure range in aluminum and to 4–8 Mbar in copper and are in good agreement with the results of dynamic experiments performed using high explosives and light-gas launching facilities.

Along with the reflection technique, the method of plasma velocity measurements by pulsed x-ray radiography is being elaborated in laser experiments. A 17 µm thick aluminum target was irradiated by a laser beam of the Shiva facility with an intensity $I = 6 \times 10^{14}$ W cm^{-2} ($E = 110$ J, $\tau = 600$ ps). In this case, one of the beams of this facility irradiated a tantalum target, which gave rise to x-rays with a characteristic energy of 1.9 keV. The motion in the field of x-ray radiation was recorded with a fast-response x-ray camera (an x-ray microscope) having a temporal resolution of 15 ps and a spatial resolution of 4.5 µm, which permitted the plasma velocity to be measured at 8×10^6 cm s^{-1}.

Mention should be made of the experiments on the generation of shock waves by short-wavelength laser radiation [31], since for these modes an increase in the amplitude pressures of the shock-compressed plasma is predicted because of an increase in the fraction of absorbed laser energy and suppression of nonthermal electrons. The laser radiation with $\lambda_0 = 0.35$ μm and an intensity of $(1–2) \times 10^{14}$ W cm^{-2} ($\tau = 700$ ps) irradiated a 25 μm thick aluminum target to generate in it a shock wave with a pressure of 10–12 Mbar. In this case, the energy absorbed by the plasma amounted to $\approx 95\%$ of the incident flux. Irradiation by long-wavelength ($\lambda_0 = 1.06$ μm) radiation with $I = 3 \times 10^{14}$ W cm^{-2} (the absorbed intensity was 1.2×10^{14} W cm^{-2}) for the same formulation of the experiment produced a pressure of about 6 Mbar.

Experimental data on the shock-wave compression of hydrogen, deuterium, and inert gases in the megabar range are presented in subsection 2.2.2 and lecture 5.

3.3 Mechanics of ultrafast deformations

High-power lasers are not only a modern physical instrument for producing high energy densities, but also a unique tool for implementing and studying ultrafast processes in condensed substance states.

The physics and mechanics of interactions depend strongly on the duration of a laser pulse. For a duration of about 1 ns and above, the main mechanism is the evaporative one (figure 3.8), while the thermomechanical mechanism is basic to the femtosecond range. In both cases, ultrahigh deformation rates are dealt with.

3.3.1 Mechanical properties in ultrafast deformations

Progress of research into the high-speed deformation, disruption, and physicochemical transformations in shock waves is largely related to the development of modern techniques for measuring wave profiles with high spatial and temporal resolutions [33, 34]. To date, a great body of experimental data has been gained about the elastoplastic and strength properties of technical metals and alloys, geological materials, ceramics, glasses, polymers and elastomers, plastic and brittle single-crystals in the microsecond and nanosecond ranges of irradiation, and considerable progress has been made in the development of methods for obtaining information about the kinetics of energy release in detonation and initiating shock waves. The experimental data are used to construct phenomenological rheological models of deformation and disruption, and macrokinetic models of physicochemical transformations, which are required for calculating explosions, high-speed impacts, and high-power pulsed radiation–substance interactions.

The objects of measurements and analysis are the shock compression wave, followed by a rarefaction wave, as well as the wave interactions in the reflection of the compression pulse from the free rear surface of the test sample. The pulse of shock compression in a plane sample is usually generated by the impact of a plate accelerated in one or other way to a velocity ranging from several hundred meters per second to several kilometers per second, and to tens of kilometers per second in record experiments. The diameter-to-thickness ratios for the striker and sample are taken to be high enough so as to ensure the one-dimensionality of the wave process throughout the measurement time. The experiments involve continuous measurements of the velocity profiles $u_{fs}(t)$ of the free sample surface, for which purpose use is made of the laser Doppler velocity meters VISAR or ORVIS with a nanosecond temporal resolution.

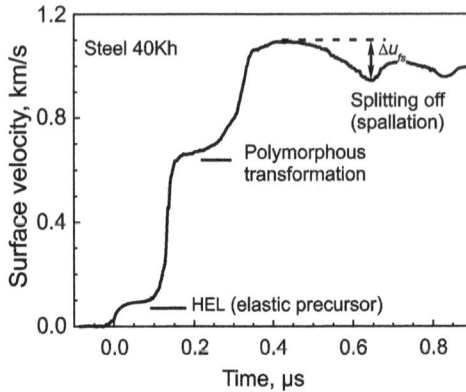

Figure 3.10. Velocity profile of the free surface of a 4 mm thick 40Kh steel sample upon an impact by a 2 mm thick aluminum plate with a velocity of 1.9 ± 0.05 km s^{-1}. Reprinted from [18] by permission from Springer. Copyright 2011.

A typical measurement result is displayed in figure 3.10, which shows the velocity profile of the free surface of a 40Kh steel sample loaded by the impact of an aluminum plate with a velocity of 1.9 ± 0.05 km s^{-1}. In this experiment, the shock compression pressure was equal to 19 GPa. The velocity profile exhibits the moments at which three compression waves sequentially reach the sample's surface. Owing to an increase in longitudinal compressibility on transiting from elastic to plastic deformation, the shock wave loses stability and splits into an elastic precursor and its following plastic compression wave. At a pressure of about 13 GPa, the material experiences a polymorphic transition from the body-centered lattice to the face-centered close-packed lattice ($\alpha \rightarrow \varepsilon$) attended with a lowering of the specific volume, with the result that the plastic compression wave splits into two in this pressure domain. The pressure behind the front of the first plastic shock wave corresponds to the onset of the transformation, while its attenuation and compression rate in the second plastic wave are determined by the kinetics of the structural transformation. After the shock wave circulation in the striker a rarefaction wave is formed, which then propagates through the sample after the shock wave. The arrival of the rarefaction wave at the sample's surface results in a decrease in the surface velocity.

The reflection of a compression pulse from the free surface gives rise to tensile stress inside the sample. The material disruption (splitting-off) under the tension is accompanied with stress relaxation and gives rise to a compression wave, which arrives at the surface in the form of a so-called split-off pulse to increase its velocity once again. Measurements of the resistance to splitting-off provide information about the strength properties of a material under submicrosecond load durations.

The longitudinal stress at the front of the elastic precursor or the Hugoniot elastic limit (HEL) is given by

$$\sigma_{\mathrm{HEL}} = 0.5 u_{\mathrm{fs}} \rho_0 c_l,$$

where u_{fs} is the free surface velocity jump in the precursor, ρ_0 is the initial material density, and c_l is the longitudinal sound velocity in the material; the stresses are assumed to be positive. The elastic limit under a one-dimensional deformation is related to the yield strength σ_T, in the ordinary sense of the word, by the equation

$$\sigma_T = \frac{3}{2}\sigma_{\text{HEL}}\left(1 - c_b^2/c_l^2\right),$$

where $c_b = \sqrt{K/\rho}$ is the 'bulk' sound velocity and K is the compression modulus.

Since the rate of plastic deformation of crystalline bodies is determined by the density of dislocations and their velocity, which are limited in magnitude, the stress of plastic flow increases with decreasing duration of the load action. In shock wave experiments, the dependence of the flow stress on the deformation rate manifests itself, in particular, in the attenuation of the elastic precursor with propagation and in the finite parameter growth times in the shock wave.

The split-off disruption in the reflection of the shock compression pulse from the free surface of a body takes place by way of initiation and growth of numerous cracks or pores. The rate of these processes and, accordingly, the stress relaxation rate in the disruption depend on the magnitude of the acting tension and may not be arbitrarily high. This is the reason why the higher the rate of load application, the greater the split-off (spall) strength. An analysis of the split-off effects in the compression pulse reflection from the free surface of a body makes it possible to determine the rupture tension (the split-off material strength) for submicrosecond load durations from the measured profile of the free surface velocity $u_{\text{fs}}(t)$. The spall strength a_{sp} is determined by the velocity decrease $\Delta u_{\text{fs}}(t)$ from the maximum velocity ahead of the front of the split-off pulse.

In the linear (acoustic) approximation, the simplified formula for determining the spall strength has the form

$$\sigma_{\text{sp}} = \frac{1}{2}\rho_0 c_b(\Delta u_{\text{fs}} + \delta),$$

where δ is a correction for the velocity profile distortion due to the difference between the velocity (c_l) of the split-off pulse, which propagates through the tensile material, and the velocity of the plastic part of the incident unloading wave in front of it (c_b). To take into account the nonlinearity of material compressibility, the value of σ_{sp} is actually calculated using equation-of-state extrapolations to the negative-pressure domain.

To date, a considerable body of measurements of the elastoplastic and strength properties of solids has been carried out with the use of laser-driven shock waves. At present, the use of the laser-driven shock wave technique permits measurements to be performed at tension levels comparable to the limiting, or 'ideal', strength of condensed substances, which defines the upper bound for the possible resistance to rupture.

Figure 3.11 shows the normalized values $\sigma_{\text{sp}}/\sigma_{\text{id}}$ of the spall strength σ_{sp} of metal single-crystals, amorphous polymers, and liquids as functions of the deformation rate [34]. Although the measured spall strengths of these materials differ by more than two orders of magnitude, in the normalized coordinates the data scatter is smaller.

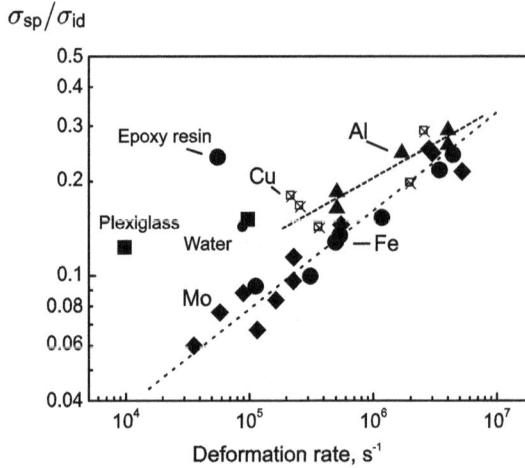

Figure 3.11. Degree of realization of the ideal split-off (spall) strength σ_{id} for homogeneous materials (single crystals, amorphous polymers, and liquids) as a function of the deformation rate. Reprinted from [18] by permission from Springer. Copyright 2011.

The data in figure 3.11 suggest that up to 30% of the ideal strength of the condensed substance is realized for nanosecond load durations. The fcc-structured plastic single-crystals of copper and aluminum exhibit a somewhat higher degree of realization of the ideal strength than iron and molybdenum, which possess a bcc crystal lattice. This is supposedly due to the possibility of higher stress densities in the neighborhood of microdefects for bcc-lattice metals with higher yield strengths. The degree of realization of the ideal spall strength for amorphous polymers and liquids is at least no smaller than for metals. The difference in the degree of realization of the ideal strength becomes smaller as the load duration becomes shorter.

Femtosecond laser techniques are employed to investigate the shock-wave phenomena in metallic films ranging from hundreds of nanometers to several micrometers in thickness. Such thin films are insufficiently rigid, and therefore experiments are carried out on metallic samples deposited by evaporation on glass substrates.

Figure 3.12 shows the results of measurements of the shock wave velocity U_S and mass velocity u_p behind the shock front in submicrometer-thick aluminum samples in comparison with its shock adiabat measured on samples several millimeters in thickness. The reason for the inconsistency of these data is that the dynamic elastic limit in submicrometer-thick aluminum samples amounts to 21 GPa, i.e. the measured values of U_S and u_p are characteristic of the shock adiabat of elastic compression. Note for comparison that the dynamic elastic limit of diamond lies in the 50–100 GPa range, i.e. in the picosecond load duration range the elastic limit of aluminum becomes comparable to the elastic limit of diamond.

The data in figure 3.13 show that the departure of the elastic shock-compressed state of aluminum realized in film samples from the three-dimensional compression shock adiabat amounts to 6.5 GPa. This corresponds to the highest shear stress of

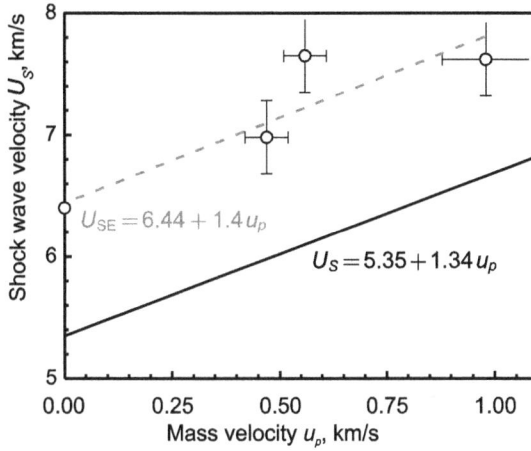

Figure 3.12. Results of measurements of the shock wave velocity and mass velocity behind the shock front in submicrometer-thick aluminum samples in comparison with its shock adiabat. Reprinted from [18] by permission from Springer. Copyright 2011.

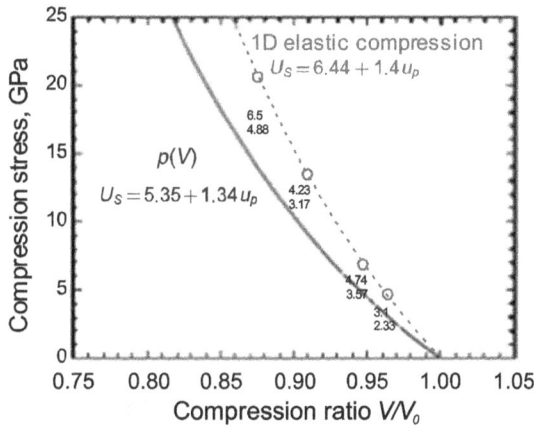

Figure 3.13. Aluminum shock-compressed states realized in submicrometer film samples irradiated by femtosecond laser pulses. Reprinted from [18] by permission from Springer. Copyright 2011.

about 4.9 GPa, which exceeds the aluminum's ideal shear strength of 3.2–3.5 GPa predicted by *ab initio* calculations and molecular-kinetic simulations. On the other hand, the elastic modulus and, accordingly, the ideal shear strength of materials increase under compression.

The experimental data on the spall strength of aluminum under shock-wave load duration ranging from tens of picoseconds to several microseconds are collected in figure 3.14. Also shown are the results of the atomistic simulation of a high-rate rupture and splitting-off and of the *ab initio* calculations of aluminum's 'ideal' strength. Resistance to split-off rupture for single crystals is higher than that for polycrystalline aluminum and aluminum alloys. Polycrystalline materials contain relatively large stress concentrators like intergranular boundaries, inclusions, etc. These defects lower the stress level required to trigger disruption. The high bulk

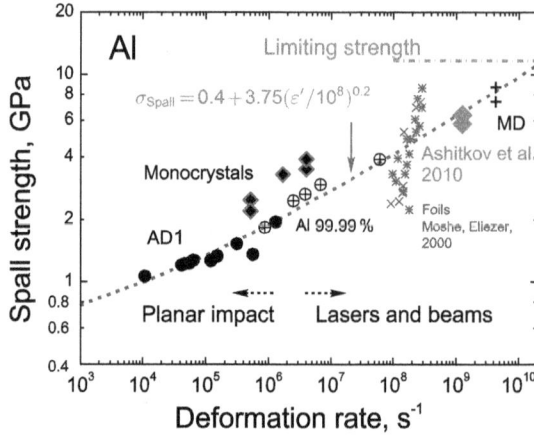

Figure 3.14. Results of split-off (spall) strength measurements for aluminum of different purity in comparison with the data for single-crystalline aluminum, molecular-dynamic splitting-off simulation data, as well as *ab initio* calculations of ideal aluminum strength. Reprinted from [18] by permission from Springer. Copyright 2011.

strength of single crystals is evidently caused by their high homogeneity. The results of measurements of the spall strength of aluminum are shown as functions of deformation rate. Here, by the deformation rate is meant the rate of substance expansion in the unloading portion of the incident compression pulse: $\dot{V}/V_0 = \dot{u}_{fs}/2c$, where $\dot{u}_{fs}$ is the decrease rate in the unloading portion of the measured wave profile. The initial disruption rate was shown to be equal to a few times this quantity. Extrapolation of the experimental data to higher deformation rates shows their agreement with molecular-dynamic calculations and predicts attainment of the 'ideal' strength for a tension rate of about 2×10^{10} s^{-1}.

3.3.2 Dynamic strength of melts and solid phases of metals

Ultrashort laser pulses provide a unique possibility of measuring the tensile strength of condensed media at extremely high tension rates $\dot{V}/V \approx 10^9$ s^{-1}. These tremendous values of the parameter $\dot{V}/V \sim u/d_T$ are obtained due to the thinness d_T of the heated target layer. For example, at a hydrodynamic expansion rate $u = 0.1$ km s^{-1} and $d_T \sim 100$ nm we obtain $\dot{V}/V \sim u/d_T \approx 10^{-3}$ ps$^{-1} = 10^9$ s^{-1}. In the calculations involving the methods of molecular dynamics, the parameter $\dot{V}/V$ is determined at precisely the point of nucleation immediately prior to its onset. The dependence of strength on the logarithm of the parameter is significant. It is shown in figure 3.15.

Experimental data (aluminum) obtained with projectile strikers and lasers are collected in figure 3.15. The data at highest rates are obtained by molecular-dynamic (MD) simulations. They find confirmation in experiments with ultrashort laser pulses (USLPs). The data of Eliezer *et al* ([35, 36], see also [37, 38]) for aluminum occupy an intermediate position between the data of MD simulations of USLP experiments, on the one hand, and the data of rather conventional experiments with the use of strikers, which pertain to microsecond times, on the other hand [34].

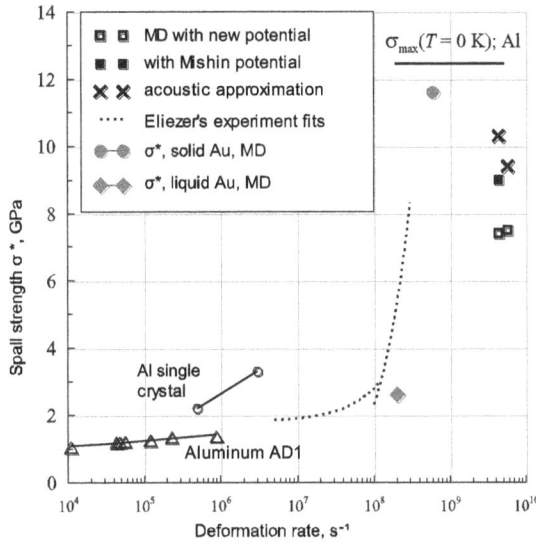

Figure 3.15. Dependence of the tensile strength σ^* on the tension rate $\dot{V}/V$. Reprinted from [18] by permission from Springer. Copyright 2011.

One group of laser experiments was performed using subnanosecond and longer-duration pulses [35–38]. The corresponding data is shown with two dashed curves [36]. Another set of data represent experiments with ultrashort laser pulses. Let us describe the data obtained in the experiments and calculations with ultrashort laser pulses. The crosses stand for the spall strength of the solid phase of aluminum, which were obtained using an estimate based on linear acoustics. This estimate is applied to molecular-dynamic simulation data. The estimate is compared with the exact MD simulation data. They are shown by empty squares.

In this case, the MD simulations were performed with a new interatomic interaction potential (EAM potential, Embedded Atom Method). The symbols in the form of empty squares pertain to the simulations with the new potential. The new potential is constructed in accordance with the data calculated by the density functional method for high-deformation states. For comparison, the full square shows the result of Mishin *et al* [39], when the EAM potential is applied in our MD simulation. Substantial uncertainties are associated with this EAM potential for high deformations of the medium. Evidently, the calculation with Mishin's potential somewhat overestimates the strength of the solid phase of aluminum.

Symbols in the form of a diamond and a solid circle in figure 3.15 pertain to MD simulations of ablation of gold in vacuum and to the splitting-off into vacuum at the rear boundary of the film irradiated from the front side by heating ultrashort laser pulses.

Figure 3.16 shows the equilibrium curve between the condensed phase and saturated vapor (binodal). This curve terminates at the critical point. The bimodal consists of two segments: the sublimation curve and the boiling curve. On the sublimation curve, the vapor pressure is negligible, and therefore in the pressure–temperature plane this curve merges with the zero-pressure axis on the linear

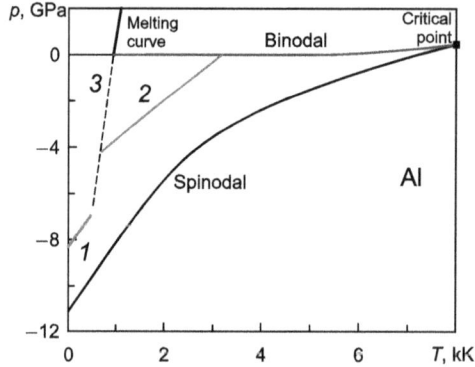

Figure 3.16. Dependence of the strength $\sigma^*(T)$ of the solid phase (curve 1) and the melt (curve 2) of aluminum on temperature. The split-off (spall) data was obtained by molecular dynamic simulation of USLP irradiation in the situation shown on the left panel in figure 3.16. The tension rate is $\dot{V}/V 2 \times 10^9$ s^{-1} in the case of a liquid and $\approx 4 \times 10^9$ s^{-1} in the case of a crystal. The crystal was stretched in direction 110.

pressure scale adopted in figure 3.16. Even at the critical point, the vapor pressure is small in comparison with the amplitudes of tension at the threshold of ablation or spallation in experiments with ultrashort laser pulses, when these amplitudes are quite high. These amplitudes approach the maximal values $p_{\text{max|cold}}$ of the modulus of negative pressure on cold crystal curves. According to the calculations with the EAM interatomic interaction potential, the $p_{\text{max|cold}}$ and $V_{\text{max|cold}}$ V_0 values amount to 12.2 and 1.5 GPa during crystal tension along direction 100, where V_0 is the equilibrium crystal volume at zero temperature.

The data in figure 3.16 indicates a sharp decrease in the strength $\sigma^*(T)$ in the vicinity of the continuation of the melting curve. Dashed curve 3 in figure 3.16 corresponds to the expected course of the melting curve in the nonequilibrium tension negative-pressure domain. It is assumed that in the case of aluminum curve 3 at low temperatures $T \to 0$ passes above the spinodel. It is believed that this differerentiates aluminum from other metals in which curve 3 intersects the spinodel at a finite temperature. In our calculations, we combine hydrodynamics and molecular dynamics (combined approach). The calculated values of pressure and temperature on curves 1 and 2 (figure 3.16) were obtained directly at the point of future fragmentation during splitting-off or nucleation with cavitation in the melt. They are determined just before the moment of fragmentation or cavitation. After this moment, the amplitude of the tensile stress at this point begins to decrease rapidly. To determine the $\sigma^*(T)$ dependence in the solid phase, it is necessary to experiment with the action of ultrashort laser pulses on solid target films.

The data in figure 3.16 pertains to aluminum. The results for gold are illustrated in figure 3.15. The strength of cold gold is higher than that of aluminum. The modulus values on the cold curves are approximately 1.8–2 times higher for gold. This is supposedly the reason why the spall strength of solid-phase gold targets is higher in our calculations simulating the actions of USLPs (see figure 3.15). In this case, aluminum in the fragmentation, which marks the onset of splitting-off in the crystal, is closer to point $p_{\text{max|cold}}$ than aluminum. For aluminum, the threshold split-off tension amounts to about 70% of the ideal strength $p_{\text{max|cold}}$; for gold the figure is 55%. We also note a substantially greater strength reduction on melting for gold in comparison with aluminum: compare the symbols corresponding to splitting-off (splitting-off in the solid phase) and ablation (cavitation in the melt) for gold in figure 3.15.

Furthermore, in experiments with ultrashort laser pulses, the tension rate $\dot{V}/V$ for gold is much lower—by about an order of magnitude—than that in the case of aluminum. This is due to the fact that, first, the heated layer thickness d_T in gold is nearly two times greater because of the far lower electron–ion thermal exchange rate. The decrease in the coefficient is mainly due to the large mass of gold nuclei. Secondly, in gold the expansion rate at the threshold of ablation or splitting-off (spallation) is several times lower due to the high density of the material.

3.4 Strong magnetic fields

Generation of superhigh-power magnetic fields in a laser plasma is one of the interesting and vigorously developing areas of research (see review [40], which we will follow in the subsequent discussion).

Different mechanisms of magnetic field generation in the interaction of high-intensity laser radiation with solid targets give rise to magnetic fields with a magnetic induction of up to 1 GG, which are produced in the interaction of high-intensity laser radiation with dense plasmas. These fields are localized near the critical surface, where the laser energy is mainly absorbed. Several main mechanisms have been proposed to generate quasi-static magnetic fields: (1) the difference in directions of the plasma temperature and density gradients; (2) the flux of fast electrons accelerated by ponderomotive forces along and across the laser pulse direction; (3) the collisionless Weibel instability.

The generation of magnetic fields with an induction of about 1 GG in relativistic dense plasmas was first predicted in [41]. According to the theory proposed in this paper, the source of a quasi-stationary magnetic field is the ponderomotive force acting on electrons. It generates a radial electron current in the direction away from the axis of the laser beam towards its periphery until a joint vibrational motion of ions and electrons due to the electroneutrality requirement begins.

The Weibel instability mechanism in the plasma arises from the anisotropy of electrons in their velocity directions. This anisotropy emerges during ionization of atoms and atomic ions by an ultrastrong laser field. Most electrons escape along the direction of the electric intensity vector of a linearly polarized laser wave. The number of electrons escaping in the transverse directions is considerably smaller. Both the longitudinal and transverse velocities are defined by the energy–time uncertainty relations. Weibel [42] was the first to show that the presence of electron current anisotropy gives rise to instability in Maxwell's equations relative to a spontaneous buildup of a quasi-static magnetic field.

The thermoelectric mechanism [40], unlike the previous one, is realized in a collisional plasma in which there are gradients of the electron density and electron temperature directed at an angle to each other. The density gradient is directed along the radius of the electron beam. It is caused by the nonuniformity of laser radiation intensity across the focal spot. As a result, the number of electrons on the laser beam axis is far greater than that at the beam periphery owing to a strong difference in the degree of ionization of the atoms of the medium. The temperature gradient is evidently directed along the normal to the target surface. The growth increment of a spontaneous magnetic field is proportional both to the temperature gradient and the velocity gradient. In this case, the magnetic field has toroidal symmetry: its circular lines of force embrace the laser beam.

In the passage of a laser pulse of relativistic intensity, plasma electrons are accelerated along the direction of laser pulse propagation by the magnetic part of the Lorentz force. This gives rise to a magnetic field, which is also annular in character.

There also exist more sophisticated methods of magnetic field generation in a laser plasma, a part of which is considered in [40].

The authors of [43] measured the magnetic field in a subcritical laser plasma. An azimuthal magnetic field with an induction of 2–8 MG on a scale of the order of 200 μm was shown to emerge for a linearly polarized laser pulse duration of 30 fs and an intensity of 4.2×10^{18} W cm^{-2}; for an intensity of 8×10^{18} W cm^{-2} the field amounted to 7 MG. Both magnetic fields exist for several picoseconds. Their generation is attributed to fast electron fluxes produced in a laser plasma.

Measurements [44] permit the range of generated magnetic fields to be determined in a dense plasma on a critical surface, which is found to correspond to 340–460 MG (figure 3.17). The magnetic pressure of such fields exceeds 10^9 atm.

The results of experiments on irradiation of solid targets by a laser pulse with an intensity of the order of 10^{20} W cm^{-2} are presented in [45]. Polarization measurements of radiation harmonic yields made it possible to find that magnetic fields with an induction of about 700 MG arise in the laser plasma. These magnetic fields also exist in the domains that far exceed the skin layer. This is attributable to the ultrarelativistic motion of electrons in so high a laser field as well as to the deep penetration of the high-order harmonics of laser radiation into the target substance.

In [46], the magnetic field induction in laser experiments was estimated at about 1 GG.

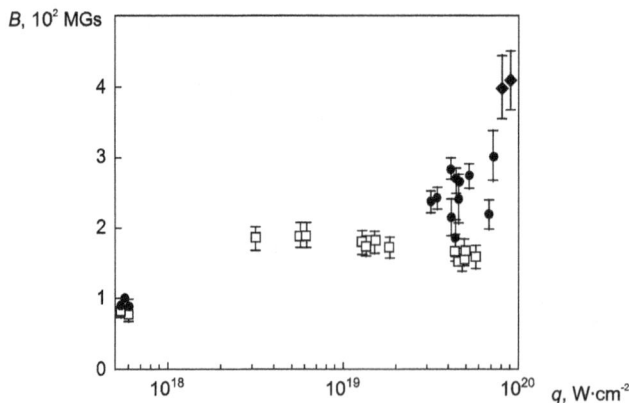

Figure 3.17. Measured magnetic field induction as a function of laser radiation intensity [44] at 5ω (♦), 4ω (•), and 3ω (□) harmonics. Reprinted from [18] by permission from Springer. Copyright 2011.

References

[1] Anisimov S, Imas Y A, Romanov G S and Khodyko Y V 1970 *Deistvie Izlucheniya Bol'shoi Moshchnosti na Metally (Action of High-Power Radiation on Metals)* (Moscow: Nauka)

[2] Anisimov S I, Ivanov M F and Inogamov N A *et al* 1977 Chislennoe Modelirovanie Lazernogo Nagrevaniya i Szhatiya Prostykh Obolochechnykh Mishenei (Numerical simulation of laser-driven heating and compression of simple shell targets) *Fiz. Plazmy* **3** 723–32

[3] Anisimov S I, Prokhorov A M and Fortov V E 1984 Application of high-power lasers to study matter at ultrahigh pressures *Sov. Phys.-Usp.* **27** 181–205

[4] Arkani-Hamed N, Dimopoulos S and Dvali G 1999 Phenomenology, astrophysics, and cosmology of theories with submillimeter dimensions and TeV scale quantum gravity *Phys. Rev.* D **59** 086004

[5] Atzeni S and Meyer-ter-Vehn J 2004 *The Physics of Inertial Fusion* (Oxford: Oxford University Press)

[6] Avrorin E N, Simonenko V A and Shibarshov L I 2006 Physics research during nuclear explosions *Phys.-Usp.* **49** 432

[7] Avrorin E N, Vodolaga B K, Simonenko V A and Fortov V E 1993 Intense shock waves and extreme states of matter *Phys.-Usp.* **36** 337–64

[8] Batani D, Vovchenko V I and Kanel' G I *et al* 2003 Mekhanicheskie Svoistva Veshchestva pri Bol'shikh Skorostyakh Deformirovaniya, Vyzvannogo Deistviem Lazernoi Udarnoi Volny (Mechanical properties of a material at ultrahigh strain rates induced by a laser shock wave) *Dokl. Akad. Nauk.* **48** 123

[9] Belyaev V S, Krainov V P, Lisitsa V S and Matafonov A P 2008 Generation of fast charged particles and superstrong magnetic fields in the interaction of ultrashort high-intensity laser pulses with solid targets *Phys.-Usp.* **51** 793

[10] Benuzzi-Mounaix A, Koenig M and Ravasio A *et al* 2006 Laser-driven shock waves for the study of extreme matter states *Plasma Phys. Control. Fusion* **48** B347–58

[11] Billon D, Cognard D and Launspach J *et al* 1975 Experimental study of plane and cylindrical laser driven, shock wave propagation *Opt. Commun.* **15** 108–11

[12] Didenko A, Rashchikov V and Fortov V 2011 On possibility of high-power terahertz emission from target under the action of powerful laser pulses *Tech. Phys. Lett.* **37** 256–8

[13] Eliezer S, Mendonca J T, Bingham R and Norreys P 2005 A new diagnostic for very high magnetic fields in expanding plasmas *Phys. Lett.* A **336** 390–5

[14] Eliezer S, Moshe E and Eliezer D 2002 Laser-induced tension to measure the ultimate strength of metals related to the equation of state *Laser Part. Beams* **20** 87–92

[15] Fortov V E 2009 Extreme states of matter on Earth and in space *Phys. Usp.* **52** 615–47

[16] Fortov V E 2011 *Extreme States of Matter. Series: The Frontiers Collection* (Berlin: Springer)

[17] Fortov V E, Batani D and Kilpio A V *et al* 2002 The spall strength limit of matter at ultrahigh strain rates induced by laser shock waves *Laser Part. Beams* **20** 317–20

[18] Garnov S V and Shcherbakov I A 2011 Laser methods for generating megavolt terahertz pulses *Phys. Usp.* **54** 91–6

[19] Ginzburg V L 1995 *O Fizike i Astrofizike (About Physics and Astrophysics)* (Moscow: Byuro Kvantum)

[20] Henderson D (ed) 2003 *Frontiers in High Energy Density Physics* (Washington: National Research Council, Nat. Acad. Press)

[21] Kanel G I, Fortov V E and Razorenov S V 2007 Shock waves in condensed-state physics *Phys.-Usp.* **50** 771–91

[22] Kanel G I, Razorenov S V, Utkin A V and Fortov V E 1996 *Udarno-Volnovye Yavleniya v Kondensirovannykh Sredakh (Shock-Wave Phenomena in Condensed Media)* (Moscow: Yanus-K)

[23] Kruer W L 1988 *The Physics of Laser Plasma Interaction* (Reading MA: Addison-Wesley)

[24] Kuramitsu Y, Sakawa Y and Dono S *et al* 2012 Kelvin–Helmholtz turbulence associated with collisionless shocks in laser produced plasmas *Phys. Rev. Lett.* **108** 195004

[25] Mishin Y, Farkas D, Mehl M J and Papaconstantopoulos D A 1999 Interatomic potentials for monoatomic metals from experimental data and *ab initio* calculations *Phys. Rev.* B **59** 3393–407

[26] Moshe E, Dekel E, Henis Z and Eliezer S 1996 Development of an optically recording velocity interferometer system for laser-induced shock waves measurements *Appl. Phys. Lett.* **69** 1379–81

[27] Mourou G A, Barry C P J and Perry M D 1998 Ultrahigh-intensity lasers: physics of the extreme on a tabletop *Phys. Today* **51** 22–8

[28] Najmudin Z, Walton B R and Mangles S P D *et al* 2006 Measurements of magnetic fields generated in underdense plasmas by intense lasers *AIP Conf. Proc.* **827** 53–64

[29] Prokhorov A M, Anisimov S I and Pashinin P P 1976 Laser thermonuclear fusion *Sov. Phys.-Usp.* **19** 547–60

[30] Pukhov A 2003 Strong field interaction of laser radiation *Rep. Prog. Phys.* **66** 47–101

[31] Ross J S, Glenzer S H and Amendt P *et al* 2012 Characterizing counter-streaming interpenetrating plasmas relevant to astrophysical collisionless shocks *Phys. Plasmas* **19** 056501

[32] Sarkisov G S, Bychenkov V Y and Novikov V N *et al* 1999 Self-focusing, channel formation, and high-energy ion generation in interaction of an intense short laser pulse with a he jet *Phys. Rev.* E **59** 7042–54

[33] Sudan R N 1993 Mechanism for the generation of 10^9g magnetic fields in the interaction of ultraintense short laser pulse with an overdense plasma target *Phys. Rev. Lett.* **70** 3075–8

[34] Tan K O, James D J and Nilson J A *et al* 1980 Compact 0.1 TW CO_2 laser system *Rev. Sci. Instrum.* **51** 776–80

[35] Tatarakis M, Gopal A and Watts I *et al* 2002 Measurements of ultrastrong magnetic fields during relativistic laser–plasma interactions *Phys. Plasmas* **9** 2244–50

[36] Trainor R J, Holmes N C and Anderson R A 1982 *Shock Waves in Condensed Matter-1981* ed W J Nellis, L Seaman and R A Graham (New York: American Institute of Physics) p 145

[37] Trainor R J, Shaner J W, Auerbach J M and Holmes N C 1979 Ultrahigh-pressure laser-driven shock-wave experiments in aluminum *Phys. Rev. Lett.* **42** 1154–7

[38] Trunin R F 1994 Shock compressibility of condensed materials in strong shock waves generated by underground nuclear explosions *Phys.-Usp.* **37** 1123

[39] Vacca J R 2004 *The World's 20 Greatest Unsolved Problems* (Upper Saddle River, NJ: Prentice Hall)

[40] Veeser L R and Solem J C 1978 Studies of laser-driven shock waves in aluminum *Phys. Rev. Lett.* **40** 391–4

[41] Veeser L R, Solem J C and Lieber A J 1979 Impedance-match experiments using laser-driven shock waves *Appl. Phys. Lett.* **35** 761

[42] Wagner U, Tatarakis M and Gopal A *et al* 2004 Laboratory measurements of 0.7 GG magnetic fields generated during high-intensity laser interactions with dense plasmas *Phys. Rev.* E **70** 026401

[43] Weibel E S 1959 Spontaneously growing transverse waves in a plasma due to an anisotropic velocity distribution *Phys. Rev. Lett.* **2** 83–4

[44] Zasov A V and Postnov K A 2006 *Obshchaya Astrofizika (General Astrophysics)* (Fryazino: Vek-2)

[45] Zigler A, Palchan T and Bruner N *et al* 2011 5.5–7.5 MeV proton generation by a moderate-intensity ultrashort-pulse laser interaction with H_2O nanowire targets *Phys. Rev. Lett.* **106** 134801

[46] Zimmerman G B and Kruer W L 1975 Numerical simulation of laser-initiated fusion *Comm. Plasma Phys. Contr. Fusion* **2** 51–60

Chapter 4

Lecture 4: Collisions of relativistic ions

4.1 Accelerators

The highest energy densities attainable under terrestrial conditions are obtained in relativistic heavy-ion collisions. Accelerators [1] required for this purpose operate in several laboratories throughout the world and are well known as the principal experimental tool in nuclear physics, elementary particle physics, quantum chromodynamics, and superdense nuclear matter physics [2–4], i.e. in the areas which have always been at the forefront of the natural sciences. This calls for constant advancement into the domain of higher energies and higher phase densities of accelerated particle beams.

Accelerator science and technology have come a long way from the first 1.2 MeV proton cyclotron invented by E Lawrence in 1932 to the Large Hadron Collider (LHC) (figure 4.1), built at CERN, accelerating the protons to a velocity of only a millionth of a percent lower than the speed of light and with an energy of colliding beams of 7 TeV each, which is 7000 times greater than the rest energy of the proton, $m_p c^2$. In the center-of-mass system, this corresponds to a proton collision energy of about 14 TeV.

During this period, the world has seen the construction of dozens of accelerators of different types, which are giant electrical facilities incorporating cutting-edge engineering ideas and exhibiting a high degree of reliability. Today, they are unsurpassed record-breakers in high-energy-density physics.

The LHC accelerator complex is being employed to collide two proton beams with an energy of 7×7 TeV to reach the new domain of distances of 10^{-16} cm and energies on the 1 TeV scale, which is sufficient, in particular, for the production of the particles of dark matter (their mass $m_{DM} \approx 10$ GeV–1 TeV), the Higgs boson, quark–gluon plasma, perhaps for discovering new dimensions, and for the solution of other intriguing problems of high energy physics [5, 6].

doi:10.1088/2053-2563/ab1091ch4

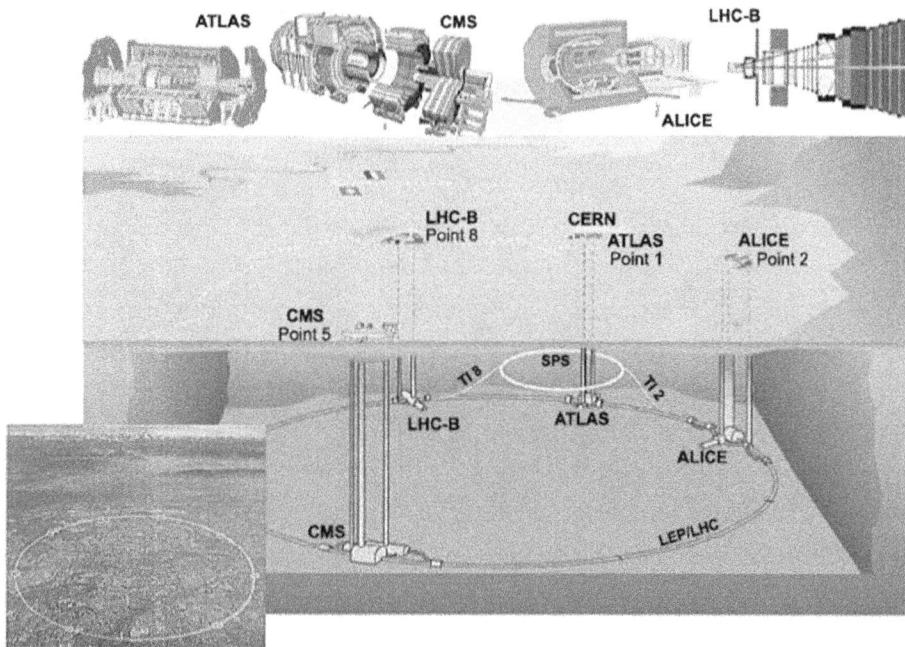

Figure 4.1. Schematic representation of the Large Hadron Collider (LHC) at the European Center for Nuclear Research (CERN). Its underground tunnel measures about 27 km in diameter. Shown at the top are the main LHC detectors: ALICE, ATLAS, CMS, and LHC-B. Reprinted from [12] by permission from Springer. Copyright 2011.

The main goal of these experiments [7] is to reveal the mechanisms of violation of electroweak symmetry by recording the Higgs boson and other new particles associated with the possible expansion of the Standard Model.

The four giant detectors—the largest one having a volume half that of the Cathedral of Notre-Dame de Paris and the heaviest one containing more iron than the Eiffel Tower—will measure the parameters of the thousands of particles that fly apart at each collision event. Despite the huge size of the detectors, the individual elements must be assembled with an accuracy of 50 μm. Later on, the detectors will be used to study the processes in the collision of highly ionized lead ions ($Pb^{82}+$) with energies of up to 155 GeV per nucleon. The LHC project was discussed in detail in subsection 2.4.1.

The Stanford Linear Accelerator (USA) generates a 5 ps pulse of 10 electrons with a kinetic energy of 50 GeV, which is focused to a spot 3 μm in size to provide a power density of 10^2 W cm^{-2}.

The actively operating Relativistic (99.99% of the speed of light) Heavy Ion Collider (RHIC) (figure 4.2) at Brookhaven National Laboratory (USA) provides a center-of-mass energy of 100–500 GeV per nucleon for colliding gold ions, 39 TeV for Au + Au, and 13 TeV for Cu + Cu [8]. In head-on collisions, 5000 elementary particles are produced. Only a few of them carry the necessary information. The new experimental data obtained on this accelerator are discussed in [8, 9]. In November 2007 in Darmstadt, Germany, the construction of a unique facility for antiproton

Figure 4.2. Relativistic Heavy Ion Collider (RHIC) of Brookhaven National Laboratory. Reprinted from [12] by permission from Springer. Copyright 2011.

and ion research (FAIR) was started to provide an energy of 1.5–34 GeV per nucleon for $\approx 5 \times 10^{11}$ and 4×10^{13} accelerated U^{92+} ions and antiprotons, respectively.

The construction cost of each of these largest ultra-relativistic hadron accelerator complexes (LHC and RHIC) is several billion dollars and is close to the limit of the economic potentials for the world's richest countries and even such an international community as the European Union.

The scientific programs at these complexes involve experimental research into the basic problems of high energy physics in hadron collisions, which are accompanied by the production of superdense nuclear substance, i.e. a quark–gluon plasma. In accordance with modern concepts, this was precisely the state of the Universe's substance during the first microseconds after the Big Bang, and also the state of the substance of such astrophysical objects as gamma-ray bursts, neutron stars, and black holes.

The problem of increasing the colliders' energy makes topical the further experimental investigations to improve the accuracy of measurements without raising the particle energy. Examples are the measurements of the magnetic moments of an electron and a muon, atomic energy level shifts, mass mixing in K^0-, D^0-, and B^0-mesons, rare decays, etc. In particular, when measuring the magnetic moment of a muon, a deviation from the values predicted by the Standard Model was obtained at a confidence level of 3.2σ only in the ninth decimal place.

For our consideration, it is important that these acceleration experiments are aimed at producing particle beams of ultrarelativistic energies to investigate not only individual hadron collision events, but also macroscopic heating of substances [1, 10].

4.2 Production of macroscopic hot plasma volumes

The methods of gas-dynamic acceleration of condensed strikers, as described in subsections 2.2.2 and 2.2.3, have a fundamental disadvantage arising from the limited value of the speed of sound c_s in the pushing gas, as a result of which the acceleration efficiency sharply (exponentially) decreases when the accelerated striker reaches the speed of sound. The high-energy-density generation techniques relying on the use of high-intensity fluxes of charged particles—electrons, light or heavy ions—as well as electrodynamic acceleration techniques, where the speed of light fulfills the role of the speed of sound, are devoid of these limitations. An important positive feature of beams of charged particles is the volume character of their energy release [10]. This distinguishes them from laser irradiation, where the main energy release of radiation with frequency ω_l occurs in a narrow critical zone, $\omega_l \sim \omega_p \sim \sqrt{4\pi e^2 n_e m_e}$, and then is transferred to the target interior by electronic thermal conductivity [11–14].

As a result of deceleration of the charged particles, there emerges a layer of isochorically-heated plasma, which subsequently expands to generate a shock wave traveling into the target or to produce a cylindrical shock wave diverging from the beam axis. Modern research into the area of high-energy-density physics takes advantage of both of these techniques, i.e. isochoric heating and compression by shock waves generated by corpuscular beams.

Either cyclotrons developed for the study of high-energy physics and nuclear physics, or high-current diode systems are used as generators of corpuscular beams [15]. In the latter case, we are dealing with subnanosecond current pulses of the megampere range with a kinetic particle energy of 1–20 MeV.

Relativistic electron beams with an energy of the order of MeV were used to excite shock waves in aluminum targets in order to study the features of electron absorption in a dense plasma and to elucidate the effect of the intrinsic magnetic fields of the beam on its bremsstrahlung in a magnetized plasma (magnetic stopping effect).

Owing to the substantially shorter paths of ions in comparison with electrons (figure 4.3 [10]), ion beams make it possible to obtain higher energy densities in plasma (figure 4.4 [10]). The high-current pulsed accelerator KALIF generated a power density on target of $\sim 10^{12}$ W cm^{-2} in a proton beam with an energy of about 2 MeV and a current of the order of 400 kA. This made it possible to accelerate thin (50–100 μm) strikers to velocities of 12–14 km s^{-1}, to perform informative measurements of the stopping power of fast protons in a dense plasma, to record the thermodynamic parameters and viscosity of shock-compressed plasma and to determine the split-off strength of metals at record-high deformation rates. It turned out (figure 4.5 [16]), for example, that the split-off strength of metals noteworthily increases (by one or two orders) with increasing deformation rate to approach its theoretical limit, which is related to the propagation kinetics of dislocations and cracks in the field of pulsed stresses [16].

Previously, the high-current pulsed BPFA-II accelerator (Sandia, USA) was used to generate megaelectronvolt beams of lithium ions with a power density of $\sim 10^{12}$ W cm^{-2}

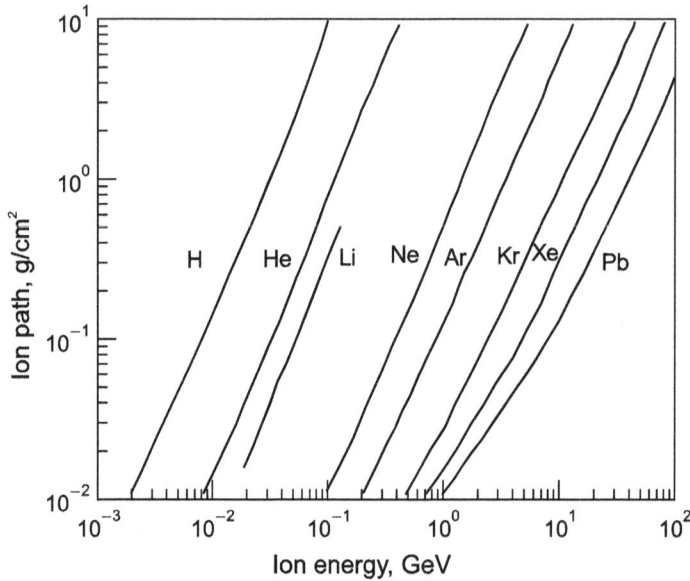

Figure 4.3. Characteristic ion paths in a cold material. Reprinted from [12] by permission from Springer. Copyright 2011.

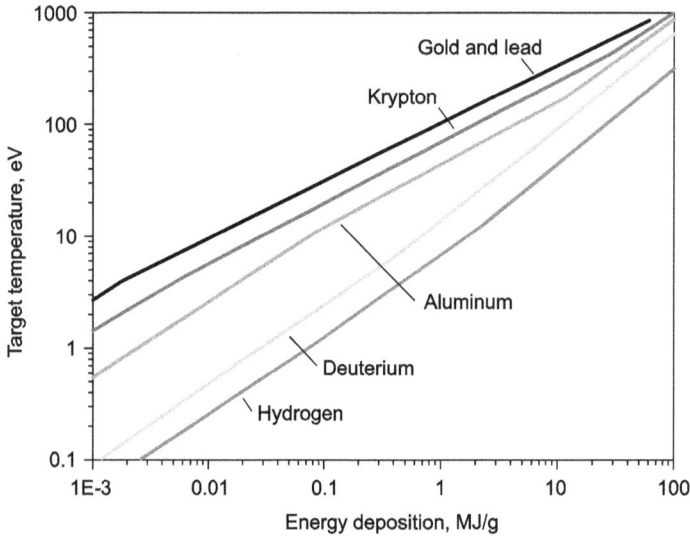

Figure 4.4. Typical plasma temperatures as a function of energy release by charged particle beams. Reprinted from [12] by permission from Springer. Copyright 2011.

for the inertial confinement fusion program. Now this setup is successfully used in the mode of a high-current Z-pinch for the production of soft x-rays and electrodynamic generation of shock and adiabatic compression waves.

The relativistic heavy ion accelerators constructed for research into the physics of extreme states of matter turned out to be promising candidates (table 2.2) for

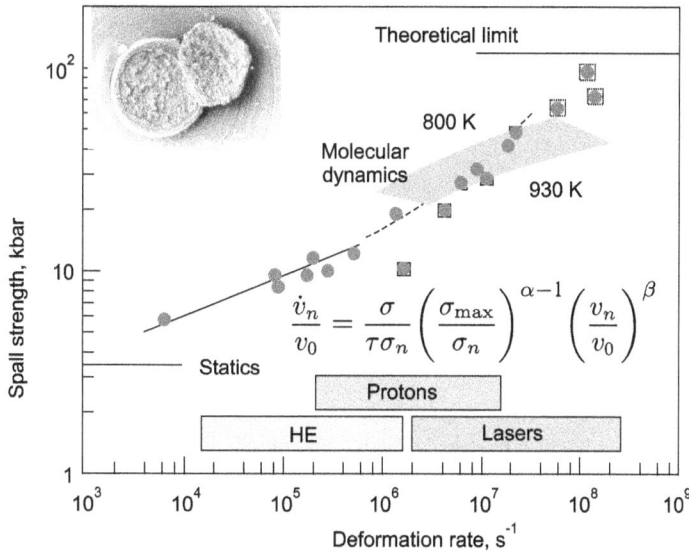

Figure 4.5. Spallation strength of an aluminum–magnesium alloy at high deformation rates. Reprinted from [12] by permission from Springer. Copyright 2011.

controlled nuclear fusion with inertial plasma confinement and for experiments on the compression and heating of dense plasma [1, 10].

The Large Hadron Collider (see figure 4.1), constructed to study the collisions of two proton beams with an energy of 7 TeV each, generates 2808 bunches of 0.5 ns duration with 1.1×10^{11} protons in each bunch spaced at 25 ns, so that the total beam duration is 89 μs, and the energy is 350 MJ, sufficient for evaporation of 500 kg of copper. The energy density in one beam is 10^{10} J cm^{-3}. The characteristic kinetic energy of one heavy relativistic ion is comparable to that of a metal liner accelerated by the explosive products of the launching system described in subsection 2.2.3.

Heavy-ion beams with a kinetic energy of 3–300 MeV/nucleon were used in experiments on heating condensed and porous targets, measuring the stopping power of ions in a plasma, and also on the interaction of charged beams with shock-compressed plasma obtained with miniexplosive-driven shock tubes [1].

Of particular interest is the use of the heavy ion accelerator at GSI in combination with the petawatt high-energy laser for heavy ion experiments (PHELIX) (see figure 2.37), which qualitatively extends the experimental capabilities of such a device.

The potential of and prospects for application of accelerators at GSI in Darmstadt (Germany) are shown in figures 2.36 and 4.6 [1]. One can see that high-intensity relativistic heavy-ion beams are interesting candidates for the generation of high-energy-density plasmas as well as for pulsed nuclear fusion [1] (see subsection 2.5.3).

4.3 Quark–gluon plasma

Given that accelerator facilities have provided a plethora of interesting physical results, we shall discuss here the generation of quark–gluon plasma (QGP) arising in

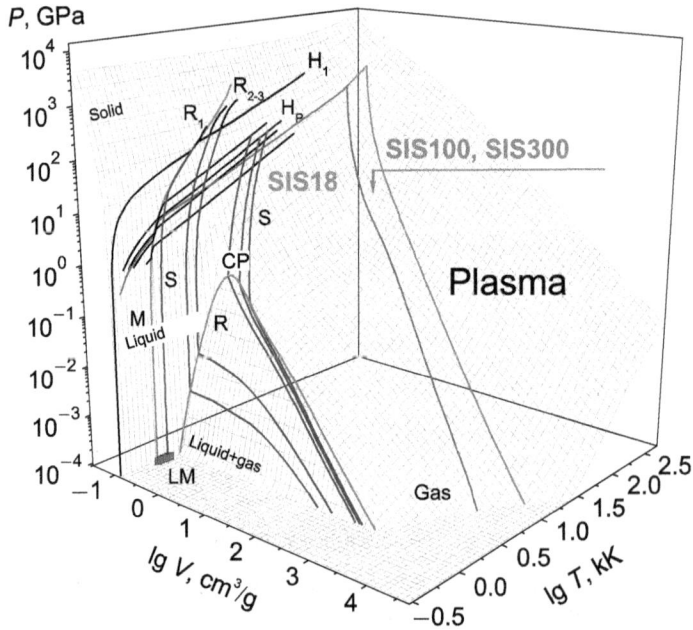

Figure 4.6. Parts of the phase diagram of zinc, attainable with heavy-ion generators. Reprinted from [12] by permission from Springer. Copyright 2011.

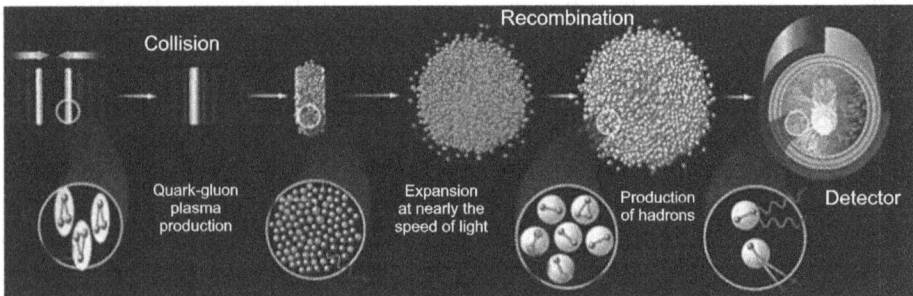

Figure 4.7. Dynamics of collisions of relativistic heavy nuclei on accelerators. Reprinted from [12] by permission from Springer. Copyright 2011.

the deconfinement of quarks at energies of no less than 200 MeV [9]. The experimental scheme is as follows: in the collision of two nuclei (figure 4.7 [8]), their kinetic energy is converted into the internal energy of the nucleons, which, in accordance with the predictions of the theory of quantum chromodynamics (QCD), leads to the appearance of the so-called 'color glass condensate (CGC)', and then, after subsequent thermalization, to the formation of a new state of matter, i.e. a quark–gluon plasma or 'quark soup' [9] (figure 4.8). Under ordinary conditions (on the left), quarks (colored balls) combine to form hadrons. At temperatures $T > T_c$, quarks deconfine, cease to be bound in hadrons and form a QGP.

The QGP arising in such collisions consists of quarks, antiquarks, and gluons [4, 17, 18]. The masses of quarks and other fermions are shown in figure 4.9 [18].

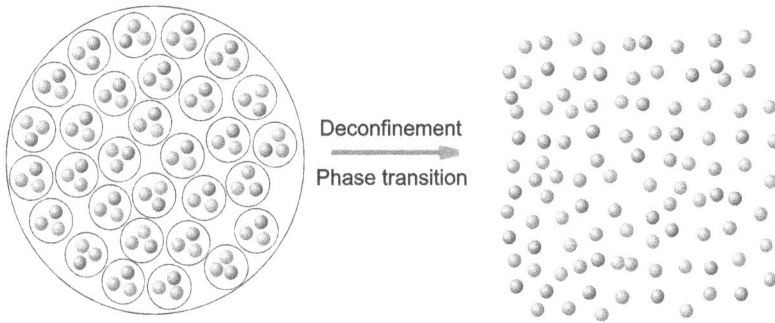

Figure 4.8. Formation of quark–gluon plasma. Reprinted from [12] by permission from Springer. Copyright 2011.

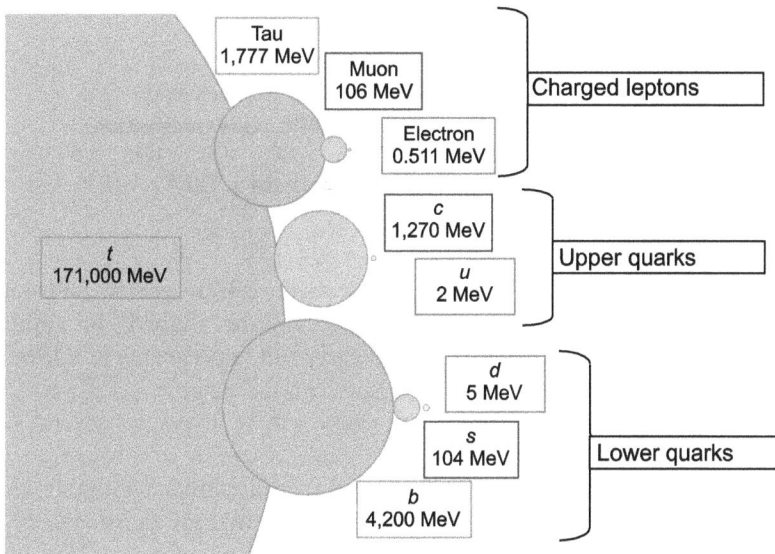

Figure 4.9. Masses of charged Standard Model fermions. The area of the circle is proportional to the mass of the particle [18]. Reprinted from [12] by permission from Springer. Copyright 2011.

This plasma is sometimes called the 'oldest' form of matter, because it existed even in the first microseconds after the Big Bang; hadrons were formed in the course of expansion and cooling of this matter. QGP has the highest density, approximately $9–10\rho_0$ ($\rho_0 = 2.5 \times 10^{14}$ g cm^{-3} is the nuclear density), and may emerge in the center of neutron stars, black holes, or in the collapse of ordinary stars (see lecture 6).

Large-scale experimental programs have been launched to study the QGP properties in collisions of ultrarelativistic ions at HERA and RHIC accelerators in Brookhaven, at the GSI accelerator facility in Darmstadt, and at the Super Proton Synchrotron (SPS) and LHC at CERN.

The first experiments with QGP in Brookhaven (RHIC) and CERN (SPS) showed more diverse behavior of such a plasma than was previously assumed (quark and gluon gas). It was found that special attention should be paid to the energy range $\sqrt{S_{NN}} \approx 2–10$ GeV, where a strongly interacting (nonideal) QGP was expected to emerge. This energy range will be studied in the framework of the CBM experiment

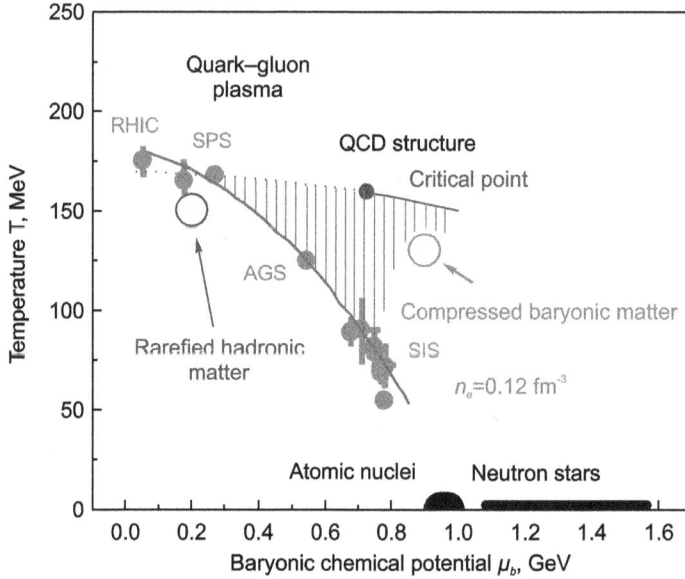

Figure 4.10. Phase diagram of nuclear matter.

of the FAIR project (see details in subsection 2.4.2), where collisions with $E_{lab} \approx 5$–35 AGeV and $\sqrt{S_{NN}} \approx 3$–8.5 GeV will be studied. In addition, the RHIC accelerator will be supposedly used to carry out experiments in a lower energy range (down to $\sqrt{S_{NN}} \approx 5$ GeV) in comparison with the range of $\sqrt{S_{NN}} \approx 200$ GeV. Unfortunately, in this case the beam luminosity will be lower by several orders of magnitude.

In any case, we are dealing with the collision of heavy nuclei with energies of the order of 100 GeV and higher per nucleus in the center of mass system or with energies of 20 TeV per nucleus in the laboratory frame of reference. The conditions attainable on modern accelerators are shown in the phase diagram of nuclear matter (figure 4.10). The region of low temperatures and baryonic densities is occupied by hadrons (nuclei and mesons) [2, 4, 17, 19]. The limiting case of high densities (5–10 times higher than the nuclear density of large nuclei: about 0.17 particle per cubic femtometer, $\approx 2.5 \times 10^{14}$ g cm^{-3}) and high temperatures ($T > 200$ MeV $\approx 10^{12}$ K) corresponds to quarks and gluons, which under these conditions are not bound in hadrons, but form a quark–gluon plasma.

The transition between these states may be either nonabrupt or abrupt, like a phase transition of the first kind with a critical point (figure 4.11 [20, 21]). To describe the behavior of compressed baryon matter in the corresponding domain of the phase diagram, use is made of the methods of quantum chromodynamics, which are also the object of experimental verification.

Like the usual electromagnetic plasma (EMP), the quark–gluon plasma can be ideal and nonideal at $T \gg T_c$ and $T \approx (1$–$3)T_c$, respectively. The corresponding nonideality parameter, i.e. the ratio between the interparticle interaction energy and the kinetic energy, has the form $\Gamma = 2Cg^2/(4\pi aT) = 1.5$–5, where C is the Casimir

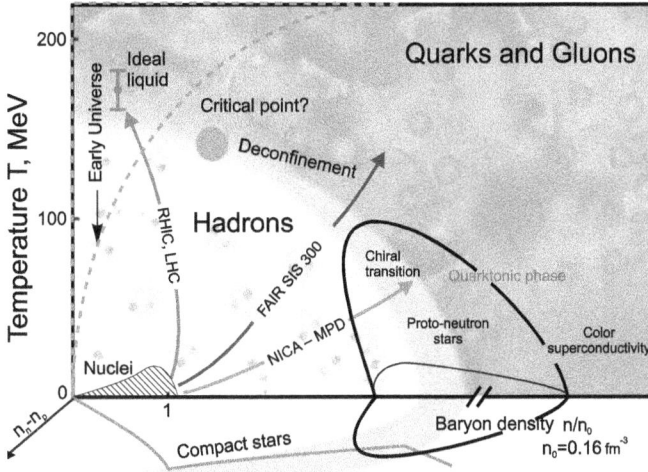

Figure 4.11. Phase diagram of strongly interacting baryonic matter (presented on the NICA website). Reprinted from [12] by permission from Springer. Copyright 2011.

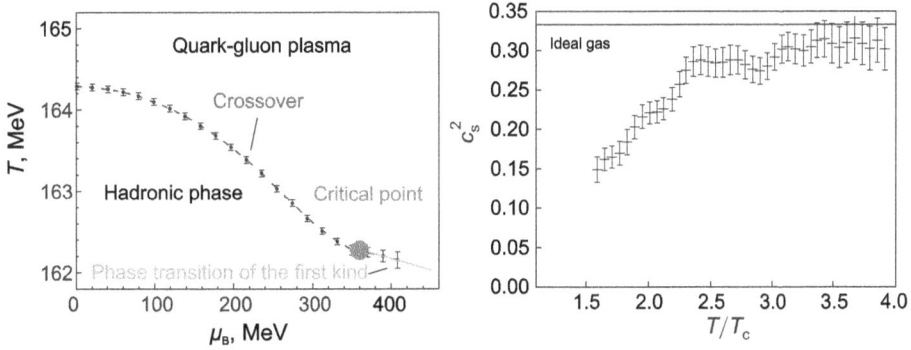

Figure 4.12. Features of the equation of state of a quark–gluon plasma [9]. Right: temperature dependence of the speed of sound. Left: phase boundary and critical point according to [22, 23].

invariant ($C = 4/3$ for quarks and $C = 3$ for gluons), $a \approx 0.5$ fm is the interparticle distance, $a \approx 1/T$; $T = 200$ eV, and $g \approx 2$ is the strong interaction constant. Factor 2 in the numerator takes into account magnetic interaction, which is of the same order of magnitude as the Coulomb interaction in the relativistic case.

At present, it is hard to tell unambiguously whether the transition to a QGP is a true thermodynamic phase transition with an energy density jump or a sharp and yet continuous transition [9].

Possibly (figures 4.12 [9, 10]), for small values of the baryon density μ_B this might be a continuous transition, and for large μ_B it is a phase transition of the first kind (figure 4.13). In any case, the theory [9] predicts a low value for the speed of sound in the transition region (figure 4.12), which is reflected in the hydrodynamic anomalies observed in the relativistic collisions of heavy nuclei. The specified features of adiabatic compressibility of the QGP testify to a 'softer' equation of state for $T \approx T_c$ and a 'stiffer' one at high temperatures as well as for $T \lesssim T_c$. In the limit $\ll T_c$, the

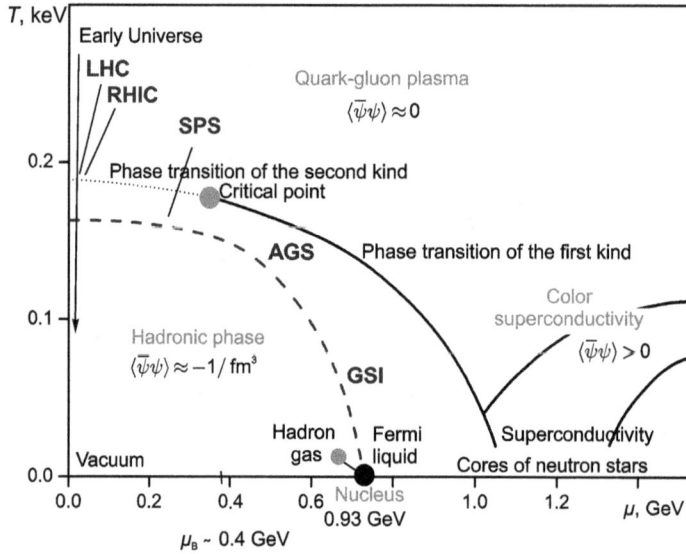

Figure 4.13. Phase diagram of QGP.

equation of state of hadronic matter becomes 'softer', although in this case, the uncertainty is quite high and the description of this matter by quantum electro-dynamics (QED) techniques is quite a challenge. Proceeding from lattice simulations in [24], a conclusion is drawn of the emergence of magnetic vortices in a QGP.

Under the conditions of the RHIC experiment, the longitudinal Lorentzian compression of the size of colliding nuclei is of the order of 100. The characteristic volume of the U + U collision domain ($\approx$3000 fm^3) contains $\approx$10 000 quarks and gluons, while the characteristic collision time is $\tau_0 \approx 0.2$–2 fm/$c \approx (5$–$50) \times 10^{-25}$ s. For this reason, part of high-energy processes supposedly takes place in the expanding substance, when the nuclear bunches have already passed through each other.

These nonequilibrium effects become stronger with increasing ion collision velocity and are a limiting factor for studying compressed baryon matter (for more details, see subsections 2.4.1–2.4.3). Thus, the characteristic atom–atom collision time, $\tau \sim 2R/\gamma$, is $\approx$1.5 fm/c for the SPS accelerator and $\approx$0.14 fm/c for the RHIC accelerator.

The authors of paper [9] drew attention to the fact that the production of fast particles in the expanding plasma after nuclear collisions is similar to the production of new forms of matter after the Big Bang. The difference is that the expansion in nuclear collisions is one-dimensional rather than three-dimensional as in cosmology.

In collisions and as the nuclear substance expands and cools down, the emergent quarks and gluons are thermalized (the time $\tau_{eq} \leqslant 1$ fm/$c \approx 3 \times 10^{-24}$ s) and may reach local thermodynamic equilibrium during the plasma lifetime $\tau_0 \approx (1$–$2)R/c \approx$ 10 fm/c. In this case, the medium will start a hydrodynamic motion; recording of this motion yields experimental information about the properties of the hadronic or quark–gluon matter, as well as the boundaries of the mutual transition, which, according to quantum electrodynamics, must occur at an energy density of the order of gigaelectronvolts per cubic femtometer.

In any case, it is believed [25] that hydrodynamics in RHIC experiments begins to work after a time period of 2×10^{-24} s, which is much less than the time of flight of the particles through the nucleus.

The idea of applying hydrodynamic equations to describe relativistic ion collisions belongs to L D Landau (1953). It turned out to be very productive for calculating a wide class of collisional experiments in a wide range of energies. In this approximation, the motion of matter at high densities is described by differential equations of conservation of mass, momentum and energy, supplemented by conservation laws for charge numbers (total baryon number, electric charge and strangeness). Use is also made of semi-empirical equations of state of nuclear matter [26].

It is evident that one of the central issues relating to the validity of hydrodynamic approximation is the question of whether the nuclear matter thermalizes under ultrarelativistic collisions.

Figure 4.14 [9] illustrates the characteristic energy densities in nuclear collisions as a function of time. Analysis of the collision and expansion dynamics shows that the transition from a relatively slow one-dimensional expansion to a faster three-dimensional expansion occurs in a characteristic time of about 0.3 fm/c. The upper part in figure 4.14 corresponds to the assumption that the system is in thermodynamic equilibrium and is an ideal massless gas, and the lower one corresponds to nonequilibrium 'frozen' conditions. By the point in time 3 fm/c, the plasma is a

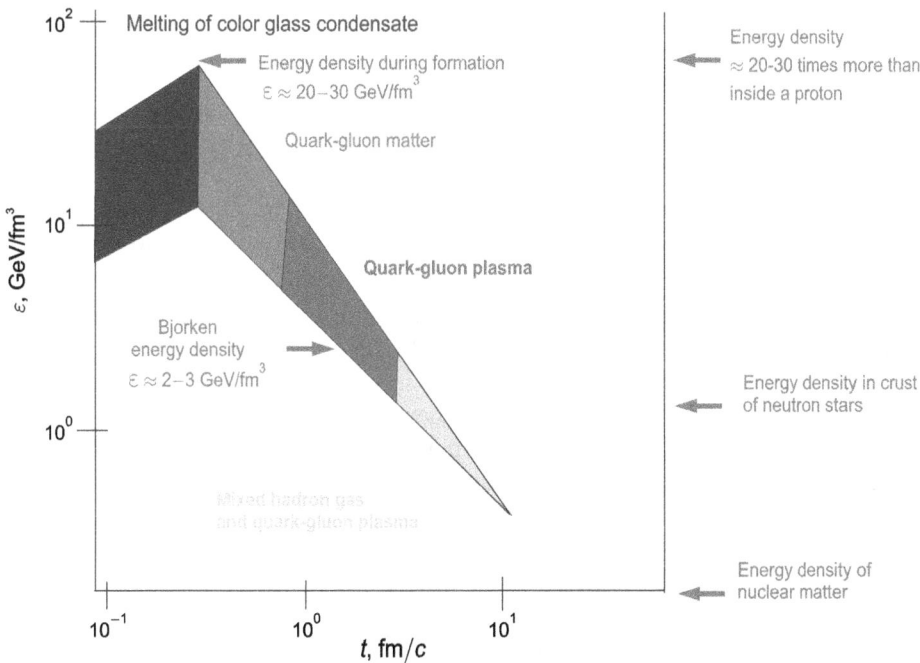

Figure 4.14. Characteristic energy densities in collisions of nuclei as a function of interaction time [9]. Reprinted from [12] by permission from Springer. Copyright 2011.

mixture of quarks, gluons, and hadrons, and at 10 fm/c, the quarks and the gluons combine to form hadrons. The lower boundary of attainable energy densities is realized for $t \approx 1$ fm/c and the upper boundary (massless gas)—for 0.3 fm/c. The general estimate of energy density has the form [9]

$$2-3 \, GeV \, fm^{-3} \leqslant E \leqslant 20-30 \, GeV \, fm^{-3}.$$

For comparison, the energy density in neutron stars (see section 6.2) amounts to ≈ 1 GeV fm^{-3}.

The emergence of new degrees of freedom in the plasma must affect the relativistic hydrodynamics of collision and expansion, which in turn is described by the equations of a viscous fluid under the conditions of local thermodynamic equilibrium.

This formalism is simplified for a nonviscous fluid (the Euler equation), while the experimental manifestation of collective (viscous) effects (see section 4.4) may be indicative of plasma effects.

The results of such comparisons for the azimuthal components of π, K, p, and Λ flow in Au + Au collisions (200 AGeV) are given in figure 4.15 [27]. One can see that simulations and measurements are in good agreement up to energies of about 1 GeV fm^{-1}, which fails at higher energies. This disagreement is attributed to the emergence of QGP. The agreement between simulations and experiments is improved by including the lowering of the speed of sound in the vicinity of $T \approx T_c$ and the corresponding 'softening' of the equation of state, which are caused by the emergence of this plasma.

The properties of the QGP can be manifested not only in the equation of state, but also in the behavior of the shear viscosity η in hydrodynamic motion:

$$T_{\mathrm{diss}}^{ij} = \eta \left(\frac{\partial v_i}{\partial x^j} + \frac{\partial v_j}{\partial x^i} - \frac{2}{3} \delta_{ij} \nabla v \right) + \xi \delta_{ij} \nabla v$$

It turns out that QGP behaves not as a gas of noninteracting particles but as a strongly interacting liquid with a vanishingly low viscosity, which, however, has a lower bound, $\frac{\eta}{S} \gtrsim \frac{1}{4\pi} \frac{\hbar}{k}$, which follows from a highly general string theory.

These effects are responsible for the underestimation of elliptical expansion velocities in central (figure 4.16 [28]) and off-center (figures 4.17 and 4.18 [29]) collisions, which were measured on the SPS, in comparison with simulations (this underestimation is unavoidable in the framework of three-dimensional (3 + 1D) hydrodynamics) and a lowering of this departure with increasing impact parameter [9], which is due to a lower efficiency of pressure transfer to the hydrodynamic flow by hadrons in comparison with plasma.

In any case, ideally nonviscous flow simulations by the Euler equations provide a better agreement with experiments (figure 4.19) than the simulations using equations with viscous dissipation.

Interestingly, for moderate energies the RHIC experiments display anomalies in QGP dissipation and yield effective viscosity values that are up to 10 times lower than one would expect from the models of weakly nonideal Debye plasma.

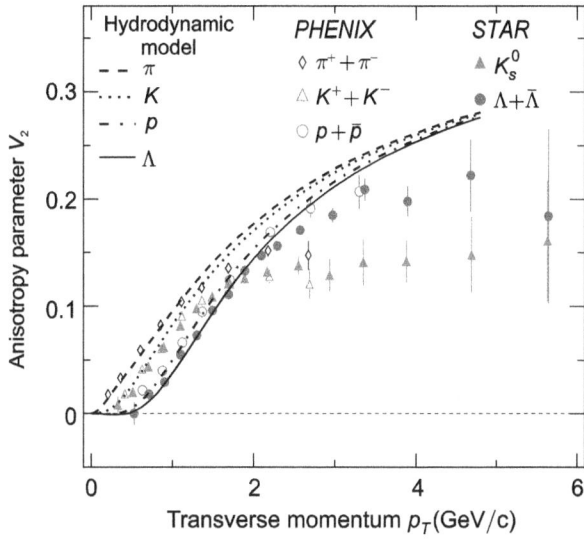

(a)

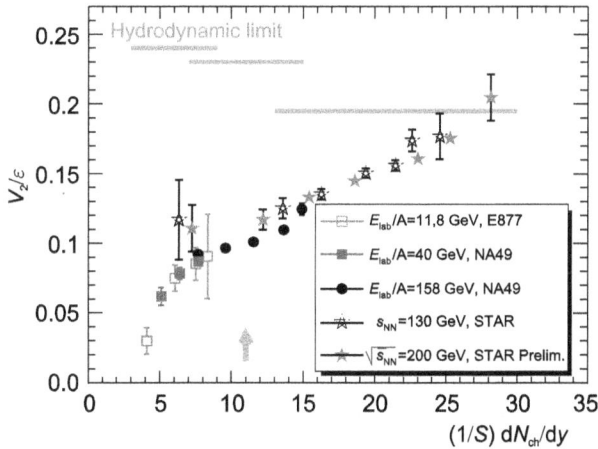

(b)

Figure 4.15. Experimental manifestation of quark–gluon plasma [9]: (a) STAR and PHENIX measurement data; (b) data from [27].

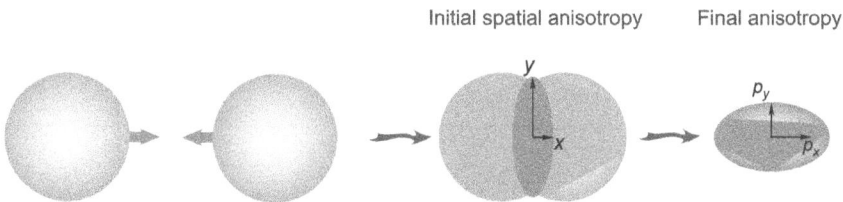

Figure 4.16. Relativistic collision of nuclei, generation of elliptical flows. The high-energy domain is elliptical in shape, so that the spatial anisotropy generates the anisotropy of momenta of the expanding medium. Reprinted from [12] by permission from Springer. Copyright 2011.

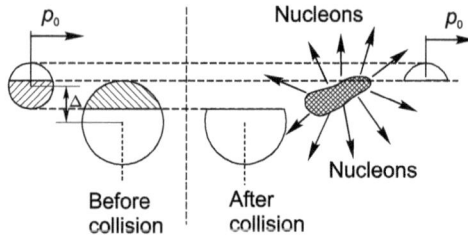

Figure 4.17. Schematic representation of off-center nuclear collisions. Reprinted from [12] by permission from Springer. Copyright 2011.

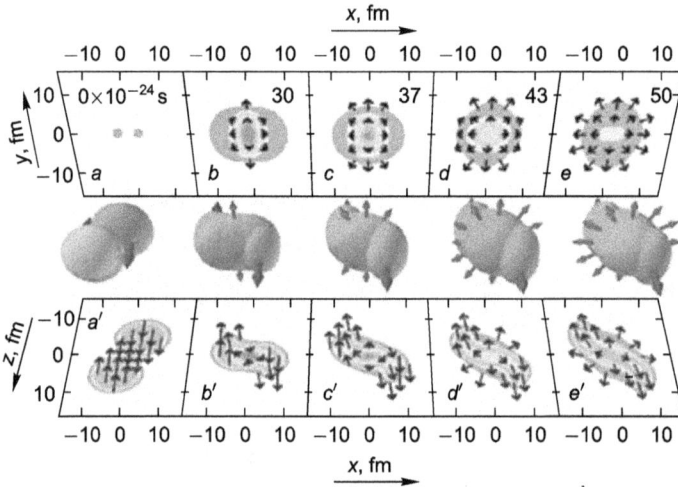

Figure 4.18. Off-center ($b = 2$ fm) Au + Au collision dynamics for 2 AGeV. The velocity of motion is plotted on the xz plane and the substance density on the xy plane. Reprinted from [12] by permission from Springer. Copyright 2011.

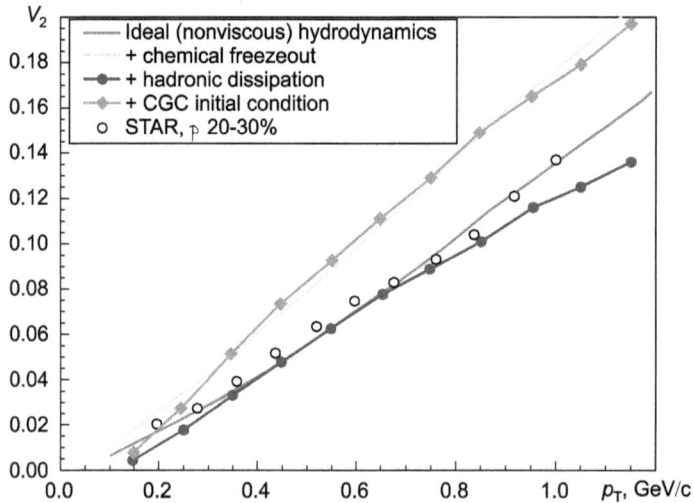

Figure 4.19. Measured (circles) and calculated expansion velocities in nuclear collisions. Reprinted from [19] by permission from Springer. Copyright 2011.

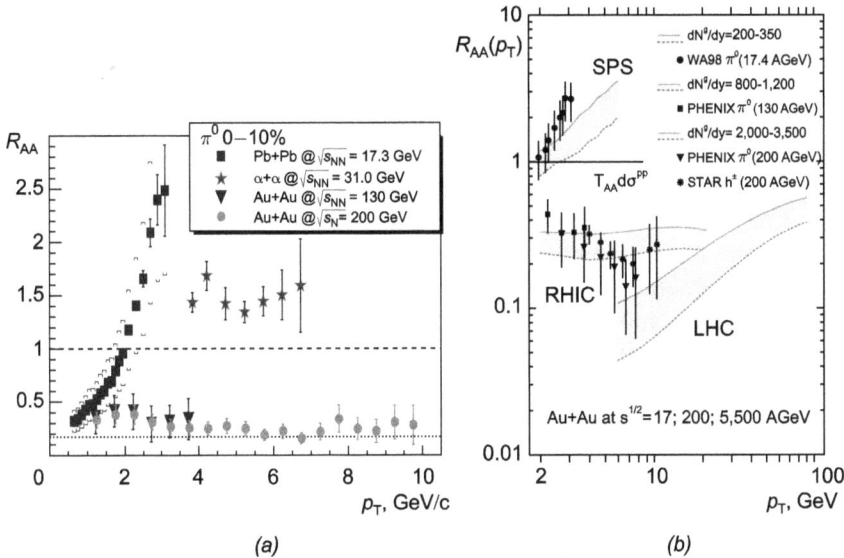

Figure 4.20. Effects of jet suppression: (a) results of suppression of π^0 fluxes recorded at PHENIX and RHIC in comparison with earlier observations at ISR and SPS, (b) measurement results for SPS, RHIC, and LHC, $R_{AA} = dN_{AA}/T_{AA}(b)d\sigma_{pp}$, in comparison with models of radiation energy losses [9]. Reprinted from [12] by permission from Springer. Copyright 2011.

Apparently [9], this is due to the effects of plasma nonideality (see figures 4.24 and 4.25).

It is amazing that the hottest and densest matter ever found in nature exceeds all known liquids in terms of the degree of nonideality, which is characterized by the viscosity of the medium. String theory and QCD calculations yield a 10 times lower viscosity value than that of superfluid helium [8]. The issue of viscosity and nonideality is highly general in nature and will be considered in the next section.

An experimental investigation of the properties of matter in the interaction domain calls for a study of the energy loss of different probe partons (quarks, gluons) that traverse it [30]. This is similar to studying the energy loss of electrons that pass through ordinary amorphous media. Recent experimental results obtained on the SPS and RHIC show clearly that the collective properties of the medium are determinative for heavy ion collisions.

The radiative properties of partons, the energy spectra of resultant hadrons, and their correlations change in comparison with those in proton–proton collisions. Induced coherent radiation reflects the collective response of the medium to the partons that penetrate it. The medium itself is characterized by the collective motion. The hadrochemical composition of the particles produced is also changed. Productive are both micro- and macroscopic approaches to the theoretical description of the observed effects in the framework of QCD. Mechanical and thermodynamic properties of the medium are studied in the framework of hydrodynamic description. Below we describe both of these approaches and discuss the corresponding experimental data.

Important information about the properties of high-density matter is obtained from the measured spectra of charged hadrons produced in collisions. For $p_T \approx$ 3 GeV, the shape of transverse-momentum charged-hadron spectra changes from an exponential dependence to a power-law one in accordance with the predictions of perturbative QCD. RHIC single-particle distribution data for central Au–Au collisions at an energy of 200 GeV demonstrate a strong p_T-independent suppression of the hadron fraction with high transverse momenta ($p_T > 4$ GeV). The measurements overlap the interval in p_T up to a value of 20 GeV. The measured large deficit of particles with a high transverse momentum is indicative of partons' energy loss in the medium. It corresponds to the effect of so-called jet suppression (quenching), which manifests itself in a softening of the hadronic spectrum obtained from in-medium partons in comparison with the spectrum in a vacuum. Therefore, the suppression factor is a powerful means for determining the density of the medium.

We see that the effects involving suppression of the jets produced in the relativistic nuclear collisions contain information about the properties of shock-compressed matter and about the emergence of QGP. By the order of magnitude this suppression is determined by the radiation loss of gluons, while the contribution of elastic loss is insignificant.

The results of such 'tomography' for PHENIX experiments are shown in figure 4.20(a) [9]; they suggest that the reduced initial gluon density must be equal to $dN^g/dy \approx 1000 \pm 200$ to explain the observed jet suppression. These values are in reasonable agreement with another set of independent measurements [9]:

(a) with the values of initial entropy determined from the plasma expansion after the collision,

(b) with the initial plasma parameters that follow from the hydrodynamic simulations of 'elliptical' flows, and

(c) with density variation rates calculated by the methods of quantum electrodynamics.

These data make it possible to find the initial energy density in relativistic collisions:

$$E_0 = E(1/\rho_0) \approx \left(\rho_0^2/\pi R^2\right) \cdot (dN^g/dy) \approx 20 \text{ GeV fm}^{-3} \approx 100 E_a$$

for the characteristic gluon momentum (1.0–1.4 GeV), which in turn defines the formation time, $\hbar/P_0 \approx 0.2$ fm/c, of the primary nonequilibrium QGP. Under these conditions, local thermodynamic equilibrium, which is necessary for application of hydrodynamics, sets in for

$$\tau_{eq} \approx (1 - 3) B/P_0 < 0.6 \text{ fm}/c.$$

By this point in time, the temperature becomes

$$T(\tau_{eq}) \approx [\varepsilon_0/(1 - 3) \times 12]^{1/4} \approx 2 T_c.$$

According to one model [9], for $P_0 \approx 2.0$–2.2 GeV the number of minijets must be of the order of 1000.

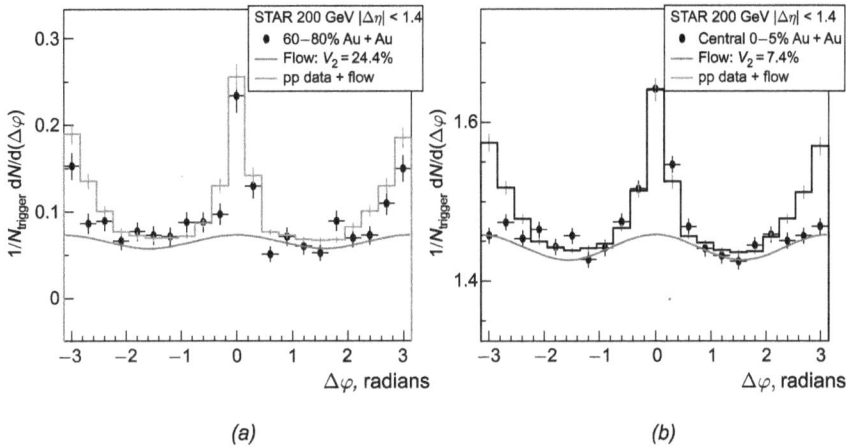

Figure 4.21. Strongly correlated backward jets in STAR and RHIC experiments [9] in Au + Au collisions (a) are compared with p + p collisions and with off-center collisions with monojets in head-on Au + Au collisions (b). Reprinted from [12] by permission from Springer. Copyright 2011.

Studying the correlation between secondary particles by recording correlated double jets in nuclear collisions yields more comprehensive data on the plasma properties (figure 4.21). It was shown [30] that both two- and three-particle correlations exhibit two oppositely directed jet-like peaks (two-jet transitions). Theoretically, to a first approximation the jets are treated as the residual manifestation of the hard scattering of quarks and gluons.

The consistency of jet characteristics obtained under different conditions (central and peripheral collisions, protons and gold) are considered [9] as a powerful argument in favor of the applicability of quantum chromodynamic techniques and QGP formation.

Therefore, the jet suppression effects observed in nuclear collisions allow the energy density of nuclear matter to be determined and conclusions to be drawn about a strong collective interaction (nonideality) of this plasma, proceeding from an analysis of the energy losses of the jets in their motion through the QGP.

Multiparticle correlations and totally (calorimetrically) reconstructed jets are the main focus of the recent efforts in the study of proton–proton and nuclear–nuclear collisions. The first results for totally reconstructed jets in proton–proton, Cu–Cu, and Au–Au collisions show clearly the jet broadening in the quark–gluon medium.

Another manifestation of QGP was the observation of entirely new event topologies in nuclear–nuclear collisions which were termed 'the ridge' and 'the double-humped event' [30]. In central collision events, the trigger jet is located on a pedestal (ridge), which spans over a broad pseudorapidity interval and falls off rapidly in the azimuthal direction. The existence of a ridge is independent of the presence of the jet peak. The characteristics of this peak coincide with the general characteristic of the particles produced, but the particle spectrum in the ridge is somewhat harder. The emergence of the ridge is unrelated to particles in jet configurations. In peripheral collisions and for high p_T of the trigger particle the ridge vanishes. Both the large pseudorapidity length of the ridge and the presence of large broad clusters revealed by two-particle correlations are indicative of the importance of collective effects.

The azimuthal $\delta\varphi = \pi$ peak in the antitrigger (remote) direction, which is observed in proton–proton collisions, is replaced with a broad antitrigger structure in Au–Au collisions. In the majority of central collision events, clearly discernable are two symmetric maxima (humps) for $\delta\varphi = \pi \pm 1.1$. The position of the peaks is virtually independent of the transverse momenta of the trigger and associated particles. The two humps supposedly merge into one broad hump at high p_T of the trigger ($6 < p_T^{\text{trig}} < 10$ GeV). This means that the antitrigger jet hidden between the two humps becomes visible again—as it must be in the case of a medium of finite size, where a parton with a high p_T escapes from the medium and forms a jet. The jets generated by heavy quarks have similar qualitative properties (though measured with a lower statistics). These features are observed both for two- and three-particle correlations. The existence of these features is undoubtedly associated with collective properties of the medium.

Among the interesting hydrodynamic phenomena, special mention should be made of Stöcker's elegant and beautiful idea [31] of the production of conic Mach shock waves (figure 4.22), the properties of which make it possible to judge the characteristics of compressed nuclear matter.

The search for and study of the physical properties of QGP are now the subject of intensive theoretical and experimental research in many laboratories around the world. These works will undoubtedly receive a new impetus after the launching of the Large Hadron Collider as a result of the implementation of FAIR and NICA projects (see subsections 2.4.2 and 2.4.3).

The quark–gluon plasma and the ordinary plasma, which makes up 98%–99% of the visible Universe and is termed 'electromagnetic plasma', have many differences and similarities. The intensity of interparticle interaction in electromagnetic plasmas is described (see lecture 2) by the nonideality parameter $\Gamma\text{EM} = \frac{z^2 e^2 n^{1/3}}{kT}$.

For quark–gluon plasma, the corresponding parameter has the form $\Gamma\text{CQP} = \frac{\alpha_S}{1-\gamma} \frac{\hbar}{lmc}$, the coupling constant (α_S) being of the order of 0.3–0.5. The mode of strong nonideality in this case corresponds to moderate temperatures $T \sim (1-2)T_c$, with a perturbative mode supposedly operating at higher temperatures.

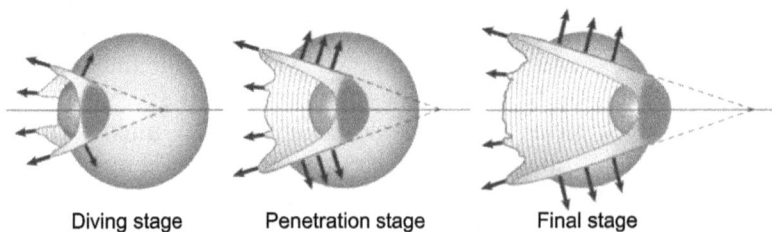

| Diving stage | Penetration stage | Final stage |

Figure 4.22. Formation of Mach shock waves in nuclear matter in the collision of a light nucleus (on the left) with a 'heavy' one. Reprinted from [12] by permission from Springer. Copyright 2011.

Unlike EMPs, the QGP is relativistic or ultrarelativistic. This difference in relativism manifests itself not only in the kinematics of motion, but also in the fact that in the nonrelativistic QGP there are only particles, the number of which is conserved, while in the relativistic one there are also antiparticles, but the number of leptons is conserved. The particle number density is no longer an adequate characteristic and is replaced by the baryon and strange particle number densities.

An increase in the fraction of various particles in comparison with a very strong suppression of pions was observed in atom–atom collisions for high transverse momenta. Therefore, the hadrochemical composition in atom–atom and proton–proton collisions is different. This is considered as a manifestation of quark–gluon plasma (or, in more general terms, as the effect of a prehadronic state). At $p_T > 2$ GeV, neutral pions and ν-mesons are strongly suppressed (approximately five-fold) in the central atom–atom collisions in comparison with proton–proton collisions.

It is noteworthy that for very small transverse momenta ($p_T < 0.5$ GeV) of the secondary particles, their fractions (different for various particles) are the same for proton–proton, atom–atom collisions, and even e^+e collisions and are independent of the initial energy, as would be expected proceeding from the universality of coherent processes for long-wavelength gluons.

In EMP, the large difference in the mass of electrons and ions leads to the difference in their dynamics and kinetics and underlies, in particular, the difference of electron and ion temperatures in relaxation processes. In QGP, there also are heavy ('charm', top, and bottom) particles, which, however, are fewer in number than the light quarks and gluons, and their lifetime is shorter. That is why the contribution of heavy quarks in QGP dynamics is small. Quark–gluon plasma is described by quantum chromodynamics, and EMP—by quantum electrodynamics. Unlike chromodynamics, the latter theory is Abelian. In quantum chromodynamics, gluons carry color charges, determining the quark–quark and quark–antiquark interaction; moreover, they interact with each other. Unlike photons, gluons make a contribution to the color charge density and the color current.

The most common feature of QGP and EMP is the collective nature of the interparticle interaction. Despite screening, the radius of effective electromagnetic interaction is usually much longer than the interparticle distance, so that there are many particles in the Debye sphere and their motion is strongly correlated. Quantum electrodynamics yields the corresponding Debye solution, $\Phi(r) = (q/\tau)e^{-m_2^D}$, in which the Debye mass (which acts as the Debye radius in the atomic system of units) $m_D^2 = e^2 T/3$ is on the order of $(qT)^2$, where q is the constant of quantum electrodynamics. Since the particle density in this theory is of the order of T^3, the number of particles in the Debye sphere is of the order of $1/q^3$ in the limit of weak compression ($1/q \gg 1$). Interestingly, the pseudopotential of the interparticle interaction of identical charges in QGP can become attractive in some cases. The long-range nature of interparticle interactions, characteristic of EMP and QGP, leads to the fact that collective effects, such as screening, plasma oscillations, instability, etc, play an important role.

Unlike experiments with EMP, where external electromagnetic or gravitational fields are used, for QGP the fields of requisite intensity are inconceivably high and only self-induced fields are of significance in relativistic collisions.

The descriptions of the EMP and QGP dynamics are also markedly different. In an electromagnetic plasma, a two-fluid (electron–ion) model with different electron and ion temperatures is widely used ($m_i \gg m_e$) under the corresponding time constraints. The condition of local electroneutrality hinders appreciable separation of charges, which leads to the equations of magnetic hydrodynamics, where the plasma moves under the action of pressure gradients and a magnetic field.

In QGP, there is no such magnetohydrodynamic analog, because each quark or gluon can carry bipolar color charges. Therefore, when local thermodynamic equilibrium is reached, different color components will have the same temperatures and velocities. In addition, the quark–gluon system becomes neutral in color even before local thermodynamic equilibrium is attained. In the case of QGP, the hydrodynamics of a neutral fluid without chromodynamic fields is realized. Of course, in the absence of local thermodynamic equilibrium, use is made of more complex kinetic equations with a collisional term of one form or another [9].

To complete the equations of motion, which express the laws of conservation of mass, momentum, and energy, it is necessary to use the equations of state. For the simplest case of an ideal ultrarelativistic gas of massless particles, it is assumed that $E_{(x)} = 3p(X)$.

Many hydrodynamic and kinetic plasma instabilities typical of EMP can be apparently manifested in QGP, although they are extremely difficult to observe in this case. However, the development of these perturbations is associated with a small (no more than 1 fm/c) measured time of the QGP thermalization and the effect of jet suppression in the relativistic collision of the nuclei.

The experimentally observed fast thermalization of matter, the parameters of elliptical flows, the spectrum of emitted particles, jet suppression, and low viscosity are attributed to the nonideality of the QGP near the deconfinement threshold. The above estimate of the nonideality parameter $\Gamma \approx 1.5$–5.0 can be an order of magnitude larger when high-order terms in the interaction potentials are taken into account. This, in turn, can cause a plasma phase transition, similar to that observed in a nonrelativistic strongly nonideal plasma [32, 33].

The plasma nonideality effects, revealed for a compressed EMP, are used to analyze the behavior of the viscosity, cross sections, collisions, and the stopping power of QGP. Since the ratio of the Landau length ($\Gamma_c \sim q^2/E$) to the Debye radius varies from one to five in QGP, this increases the scattering cross section by a factor of 2–9, which shortens the mean free path λ and, hence, decreases the viscosity ($n \sim \lambda$) by an order of magnitude (see the next section). This is consistent with the measured parameters of elliptic flows and the spectra of particles in the collision of nuclei, as well as with the recorded rise in the collision losses.

Interesting analogies arise between strongly nonideal QGP and strongly nonideal dusty plasma [33]. In both cases, we are apparently dealing with a non-Newtonian fluid, the shear viscosity of which depends on the velocity of motion. In addition, QGP possesses features of a nanofluid. Thus, the initial size of the QGP immediately after the collision is approximately 10 fm, i.e. 20 interparticle distances, which distinguishes this system from a continuous medium. This is also characteristic of nonideal dusty plasma.

In conclusion, we present a scheme for the transformation of matter at high energy densities (figure 4.23), which in a certain sense continues the conclusions of subsection 2.2.4 on the simplification of matter when moving toward extremely high pressures and temperatures.

We see that experiments on relativistic heavy ion collisions (SPS, RHIC) have become a source of invaluable information about a new field of physics, i.e. physics of the quark–gluon medium. These experiments have shown that nuclear collisions cannot be regarded as an additive superposition of proton–proton collisions; in the dynamics of heavy ions, one has to take into account the collective properties of the medium.

This is evidenced by the presence of anisotropic fluxes, jet suppression, correlations of special kind like the ridge and double-humped events as well as many other above-discussed data.

A theoretical understanding of the evolution of the medium required the full-scale involvement of QCD to discuss phenomena such as color glass condensate, glasma, thermalization, quark–gluon plasma, hadronization, and many others. Note that the methods of condensed-state physics describe a modification of the parton energy loss in matter due to effects of the formation length and the collective response of the

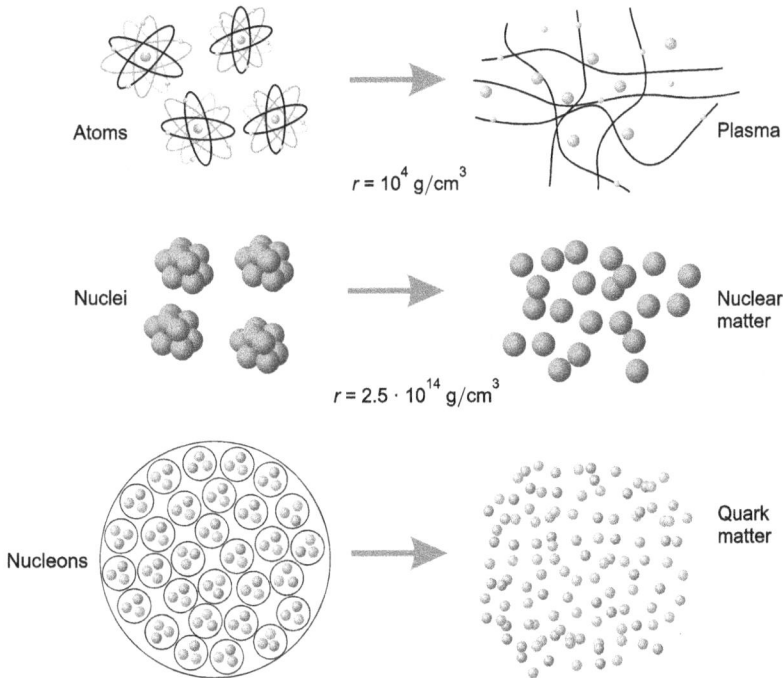

Figure 4.23. Matter transformation at high energy densities. Reprinted from [12] by permission from Springer. Copyright 2011.

medium (chromopermittivity) stemming from its polarization. Hydrodynamics was widely used to describe the collective behavior of this medium.

Many physicists look forward to reaching the next energy milestone at the Large Hadron Collider at CERN [8]. Experiments are expected to begin soon on observing collisions of lead nuclei at a total energy of more than 1 million GeV. The instantaneous energy density produced by mini-explosions in the LHC tunnel will be several times greater than in collisions at RHIC, and the temperatures will be much higher than 10^{13} K. Scientists will be able to simulate and study the conditions that existed during the very first microsecond after the Big Bang.

Extremely interesting is the question whether the similarity to the fluid found at RHIC will persist at higher temperatures and energy densities that will be obtained at the LHC [8]. Some theorists believe that the force acting between the quarks will become weak as soon as their average energy exceeds 1 GeV, and that the quark–gluon plasma will still behave like a gas. Other researchers do not share this opinion. They argue that at higher energies the QCD force does not decrease rapidly enough, and therefore quarks and gluons will remain strongly bound like molecules of a liquid.

4.4 Viscosity and interparticle interaction

As we have seen above, one of the unexpected and brilliant results obtained at RHIC accelerator is that a strongly compressed quark–gluon plasma with $p \approx 10^{30}$ bar, $T \approx 10^{12}$ K and $\rho \sim 10\rho_0 \approx 10^{15}$ g cm^{-3} behaves not as a gas of quarks and gluons but as a liquid with a vanishingly small viscosity ($\eta/s \sim (0.08–0.24)$). This is supposedly a reflection of the general basic properties of an interacting ensemble of particles, which behaves as a nonviscous ideal liquid in the limit of strong interparticle interaction [25].

This feature is inherent in a broad class of physical objects of highly different nature in an extremely broad range of parameters: 18 orders of magnitude in temperature and 25 orders of magnitude in density [34, 35]. The cases in point are a relativistic supercompressed quark–gluon plasma with $T \approx 10^{12}$ K [25], matter in the early Universe, viscosity of helium isotopes, molecular and atomic liquids, dust plasma [36], electrons in metals and semiconductors, graphene, black holes, and the Fermi gas of ultracold ($T \approx 10^{-6}$ K) lithium atoms in optical traps.

In all these classical and quantum systems, strong (collective) interparticle interaction manifests itself as a sharp decrease in shear viscosity, which serves as a diagnostic tool, performing the role of a signal indicative of the collective behavior of a strongly correlated system. It is important that this behavior of viscosity is virtually independent of the model and is 'nonperturbative.'

The concept of an ideal fluid corresponds to a macroscopic system located in local thermodynamic equilibrium and having a vanishingly small shear viscosity $\eta = \rho v^2 \tau$. From a macroscopic perspective, viscosity describes the value of the resistance of a medium to its motion. Microscopically, it is a characteristic of the interaction intensity between different elements of a medium. One can assert [25] that viscosity reflects the inability of a medium to transmit a momentum to neighboring sites. It reflects local departure from complete equilibrium, which gives rise to flow friction and deceleration.

The shear viscosity describes how perturbations propagate through a medium via interactions. The lower the viscosity of a fluid, the stronger the interactions, and the higher the intensity of perturbation transfer. In the other limiting case of an ideal gas of noninteracting particles, shear viscosity has a finite value.

The point is that viscosity is proportional to the average time between collisions of particles and the energy density of the system E. The entropy density S is proportional to the particle density N, so that $\frac{\eta}{S} \sim \frac{\tau \varepsilon}{k}$ and, where ε is the energy per particle. Because the interparticle interaction decreases τ, this leads to a decrease in η. According to the Heisenberg principle, $\eta \varepsilon \gtrsim \frac{\eta}{S}$ and $\sim \frac{\hbar}{k}$ so that the ideal fluid concentration with $\eta = 0$ contradicts quantum mechanics.

Using the arguments based on the analysis of Heisenberg's relations, gauge theory, string theory, and the holographic principle developed for the description of gravity, the authors of elegant paper [34] estimated a lower boundary of viscosity, which yielded a very small value:

$$\left(\frac{\eta}{S}\right) = \frac{1}{4\pi}\frac{\hbar}{k} \approx 6.08 \times 10^{-13} \text{ Ks.}$$

This estimate is based on the calculation of the propagation of perturbations by the methods of the gauge theory, which considers the propagation of a graviton in a multidimensional space from one point of the boundary, its reflection from a black hole in the anti-de Sitter space, and the return of a graviton back to the boundary. In this case, the graviton reflection cross section is close to the area of the black hole horizon, and the viscosity is proportional to the area of the horizon. To this end, use is made of the Hawking analogy between the black hole physics and thermodynamics. Superstring theory of gravity on a 10-dimensional 'black' membrane is considered, which has analogies in hydrodynamics. The holographic principle (Ads/CFT correspondence) reduces this multidimensional consideration to a space of smaller dimensions. In this case, the entropy of the 'black' membrane proves to be proportional to the area of the event horizon $S = \frac{A}{4G}$, from which the lower limit of viscosity follows.

The authors of paper [35] used the methods of string theory developed for strongly interacting systems to describe 'dual' weakly interacting ensembles. It was shown that η/S should have a nonzero minimum at $\frac{\eta}{S} \gtrsim \frac{1}{4\pi}\frac{\hbar}{k}$. This expression relates hydrodynamics and thermodynamics of strongly interacting systems. Specific calculations of shear viscosity of a relativistic hadron-resonance gas, as well as perturbative and lattice calculations, do not contradict this boundary estimate.

Using chiral perturbation theory [37], it was shown that the viscosity η/S decreases monotonically with increasing T (with increasing degree of correlation in the system) and reaches a minimum near the phase transition temperature of the QGP. This behavior of η/S may by itself be indicative of a phase transition in a strongly correlating system. A conclusion about the minimum of η/S in the phase transition to QGP is drawn in [38], where ultrarelativistic molecular dynamics methods were used to show that the expansion and cooling of the fireball is accompanied by an increase in η/S. The monotonous growth of this QGP parameter was obtained with increasing T.

As we saw in section 4.3, the experiment clearly shows the formation of 'elliptical' flows caused by high-pressure plasma (see figures 4.16 and 4.19) in off-center collisions that are confidently reproduced by hydrodynamic collision calculations. This is also a confirmation of the description of relativistic collisions by the methods of mechanics of a continuous medium under conditions of local thermodynamic equilibrium.

When moving in a dense QGP, the jets experience absorption (suppression), which indicates a strong opacity of the quark–gluon medium [25]. It is interesting that, since photons are not subjected to strong interaction, they are hardly absorbed by the QGP; therefore, they can carry information about the QGP temperature, which turned out to be within 300–600 MeV [25] and well above the phase transition temperature to the QGP, $T_c \approx 170$ MeV.

These data on jet suppression and elliptical flows can be described under assumption of a vanishingly small viscosity, which only slightly exceeds its lower estimate, $\frac{\eta}{S} \gtrsim \frac{1}{4\pi}\frac{\hbar}{k_B}$.

Experimental data on viscosity of strongly nonideal dusty plasmas (figure 4.24 [39]) and shock-compressed electron–ion plasmas (figure 4.25 [39]) show a decrease in η/S with increasing nonideality (correlation) in the system.

The results of collisional experiments show that they are close to the mode of weak asymptotic freedom, in which quantum chromodynamics predicts the screening of color charges at short distances.

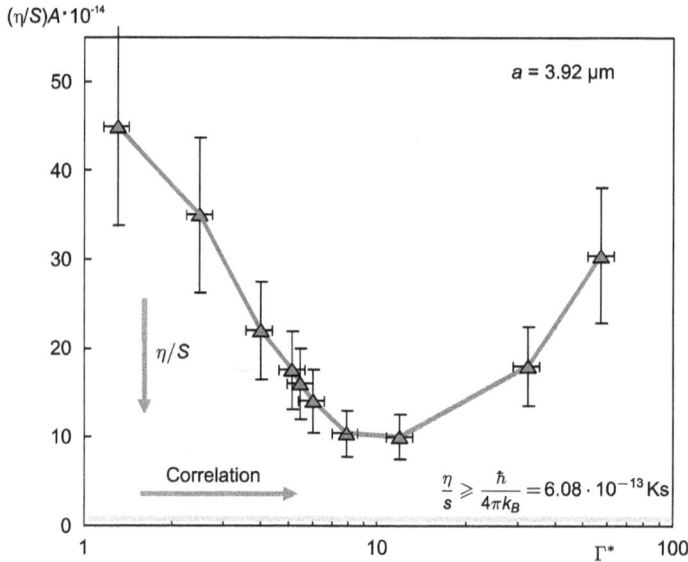

Figure 4.24. Dependence of viscosity on the nonideality parameter for dusty plasma.

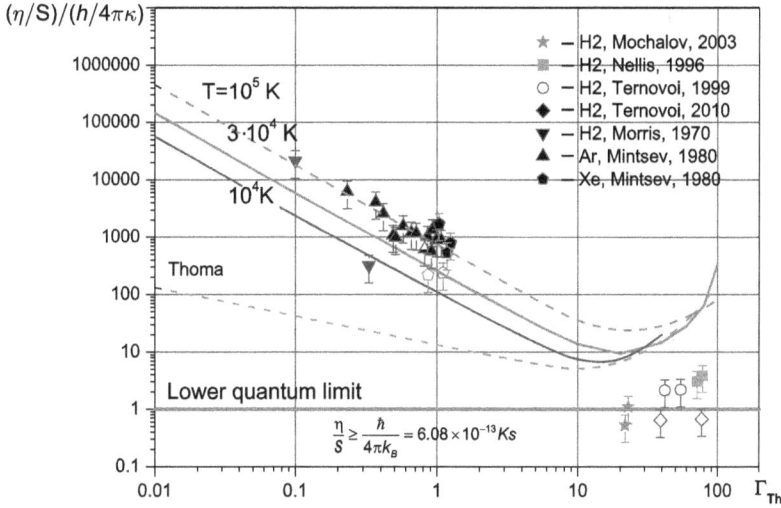

Figure 4.25. Viscosity of shock-compressed electron–ion plasma.

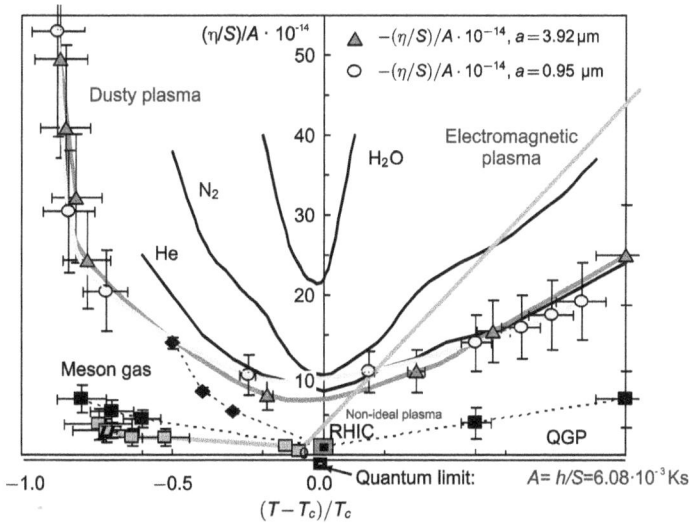

Figure 4.26. Viscosities of different media in dimensionless variables.

The analogy between the Debye screening of Coulomb charges in an electro-magnetic plasma and color screening in a quark–gluon plasma discussed in section 4.3 can be very informative.

We emphasize once again that strongly interacting systems are liable to phase transitions when the ordering interparticle interaction energy is far greater than the disordering thermal energy.

A strong collective interparticle interaction is realized in all these cases, resulting in the appearance of new phase states [33].

The quark–gluon plasma discussed above is an illustrative example of this type of phase transitions. The corresponding investigations will be conducted within the framework of the international FAIR project (see subsection 2.4.2).

To conclude this section, we present data on η/S for helium, nitrogen, water, baryonic matter in RHIC experiments, quark–gluon plasma, and meson gas as a function of reduced temperature (figure 4.26 [25]). One can see that the QGP exhibits the lowest shear viscosity.

References

[1] Alt C, Anticic T and Baatar B *et al* 2003 Directed and elliptic flow of charged pions and protons in *Pb+Pb* collisions at 4 A and 158 AGeV *Phys. Rev.* C **68** 034903

[2] Anisimov S I, Prokhorov A M and Fortov V E 1984 Application of high-power lasers to study matter at ultrahigh pressures *Sov. Phys.-Usp.* **27** 181–205

[3] Atzeni S and Meyer-ter-Vehn J 2004 *The Physics of Inertial Fusion* (Oxford: Oxford University Press)

[4] Blaschke D *et al* (ed) 2009 *Searching for a QCD Mixed Phase at the Nuclotron-Based Ion Collider fAcility (NICA White paper)*

[5] Chernodub M N, Nakamura A and Zakharov V I 2008 Manifestations of magnetic vortices in the equation of state of a Yang–Mills plasma *Phys. Rev.* D **78** 074021

[6] Cremonini S 2011 The shear viscosity to entropy ratio: a status report *Mod. Phys. Lett.* **B25** 1867–88

[7] Csikor F, Egri G and Fodor Z *et al* 2004 The QCD equation of state at finite Tμ on the lattice *Prog. Theor. Phys. Supp.* **153** 93–105

[8] Demir N and Bass S A 2009 Shear-viscosity to entropy-density ratio of a relativistic hadron gas *Phys. Rev. Lett.* **102** 172302

[9] Dremin I M and Leonidov A V 2010 The quark–gluon medium *Phys.-Usp.* **53** 1123–49

[10] Fodor Z and Katz S D 2004 Critical point of QCD at finite T and μ, lattice results for physical quark masses *J. High Energ. Phys.* **04** 050

[11] Fortov V, Iakubov I and Khrapak A 2006 *Physics of Strongly Coupled Plasma* (Oxford: Oxford University Press)

[12] Fortov V E 2013 *Extreme States of Matter. Series: The Frontiers Collection* (Berlin: Springer)

[13] Fortov V E 2007 Intense shock waves and extreme states of matter *Phys.-Usp.* **50** 333

[14] Fortov V E, Hoffmann D H H and Sharkov B Y 2008 Intense ion beams for generating extreme states of matter *Phys.-Usp.* **51** 109

[15] Fortov V E, Petrov O F, Vaulina O S and Timirkhanov R A 2012 Viscosity of a strongly coupled dust component in a weakly ionized plasma *Phys. Rev. Lett.* **109** 055002

[16] Friman B, Höhne C and Knoll J *et al* 2010 *The CBM Physics Book, Lecture Notes in Physics* vol 814 1 edn (Berlin: Springer)

[17] Ginzburg V L O 1995 *Fizike i Astrofizike (About Physics and Astrophysics)* (Moscow: Byuro Kvantum)

[18] Glendenning N 2000 *Compact Stars, Nuclear Physics, Particle Physics and General Relativity* (New York: Springer)

[19] Gyulassy M and McLerran L 2005 New forms of QCD matter discovered at RHIC *Nucl. Phys.* A **750** 30–63

[20] Henderson D (ed) 2003 *Frontiers in High Energy Density Physics* (Washington: National Research Council, Nat. Acad. Press)

[21] Jacak B and Steinberg P 2010 Creating the perfect liquid in heavy-ion collisions *Phys. Today* **63** 39–43

[22] Kanel G I, Razorenov S V, Utkin A V and Fortov V E 1996 *Shock-Wave Phenomena in Condensed Media* (Moscow: Yanus-K)

[23] Kovtun P K, Son D T and Starinets A O 2005 Viscosity in strongly interacting quantum field theories from black hole physics *Phys. Rev. Lett.* **94** 111601

[24] Kovtun P, Son D T and Starinets A O 2003 Holography and hydrodynamics: diffusion on stretched horizons *J. High Energ. Phys.* **10** 064

[25] Kruer W L 1988 *The Physics of Laser Plasma Interactions* (MA: Addison-Wesley)

[26] Lindle I 1998 *Inertial Confinement Fusion* (New York: Springer)

[27] Mangano M L 2010 QCD and the physics of hadronic collisions *Usp.-Phys.* **53** 109–32

[28] Mesyats G A 2004 *Impul'snaya Energetika i Elektronika (Pulse Power Engineering and Electronics)* (Moscow: Nauka)

[29] NICA http://theor.jinr.ru/twiki-cgi/view/NICA

[30] Novikov I D 2001 'Big Bang' echo (cosmic microwave background observations) *Phys.-Usp.* **44** 817

[31] Okun' L B 1990 *Leptony i Kvarki* 2nd edn (Moscow: Nauka)
Okun' L B 1982 *Leptons and Quarks* (Amsterdam: North-Holland) English Transl.

[32] Ollitrault J Y 1992 Anisotropy as a signature of transverse collective flow *Phys. Rev. D* **46** 229–45

[33] Riordan M and Zajc W A 2006 The first few microseconds *Sci. Am.* **294** 34A–41A

[34] Rubakov V A 2007 Hierarchies of fundamental constants (to items nos 16, 17, and 27 from Ginzburg's list) *Phys.-Usp.* **50** 390

[35] Rubakov V A 2001 Large and infinite extra dimensions *Phys.-Usp.* **44** 871

[36] Sharkov B Y (ed) 2005 *Yadernyi Sintez s Inertsionnym Uderzhaniem (Inertial Confinement Nuclear Fusion)* (Moscow: Fizmatlit)

[37] Stocker H, Hofmann J, Maruhn J and Greiner W 1980 Shock waves in nuclear matter — proof by circumstantial evidence *Prog. Part. Nucl. Phys.* **4** 133–95

[38] Troitskii S V 2012 Unsolved problems in particle physics *Phys.-Usp.* **55** 72–95

[39] Vaulina O S, Petrov O F and Fortov V E *et al* 2009 *Pylevaya Plasma (Eksperiment i Teoriya) (Dust Plasma (Experiment and Theory))* (Moscow: Fizmatlit)

IOP Publishing

Lectures on the Physics of Extreme States of Matter

Vladimir E Fortov

Chapter 5

Lecture 5: Physics of strongly compressed electromagnetic plasma

Thermodynamic properties of a compressed and heated substance in an ionized (plasma) state have always been of great interest from fundamental and applied perspectives. Dense hot plasma is the most common state of matter in the Universe—it accounts for more than 95% of the matter visible to us in stars, planets, and exoplanets [1, 2].

As the temperature of the gas rises, neutral particles dissociate. This plasma is sometimes called electromagnetic (in contrast to quark–gluon plasma, see lecture 4), emphasizing the leading role of the Coulomb interparticle interaction in it. The long-range nature of the Coulomb interaction manifests itself in a vast range of parameters of the state of matter. This feature of the Coulomb potential causes difficulties in the theoretical description [3], without permitting us to apply the usual apparatus of the statistical theory of gases to the plasma, owing to the divergence of the corresponding integrals.

The description of the thermophysical properties of plasma calls for correct consideration of strong collective interparticle interaction, degeneracy effects, correct separation of discrete and continuous energy spectra, effects of thermal and density ionization and other complex phenomena in a compressed and heated chemically reacting multicomponent medium. These phenomena determine the behavior of matter in a vast extent of the phase diagram, occupying the region from a solid and liquid to a neutral gas, encompassing the phase boundaries of melting and boiling of metals, as well as the metal–dielectric transition region. The latter problem has now received considerable attention in experiments on multiple (quasi-isentropic) shock-wave compression of dielectrics and their metallization in the megabar pressure range, on electrical explosion of conductors by pulsed currents, and also on dielectrization of strongly compressed metals [1, 2].

Today, the study of strongly compressed Coulomb systems is one of the most 'hot' and intensively developing fundamental scientific disciplines at the junction of plasma physics, physics of condensed states, atomic and molecular physics, with a wide variety of physical effects stimulated by nonideality and a constantly expanding set of objects and states where this nonideality plays a decisive role.

5.1 Hierarchy of models

As in the description of a solid and a liquid, a universal *adiabatic approximation* is used to study plasma thermodynamics. It is valid for any values of density and temperature with an accuracy proportional to the ratio of the electron mass m_e to the nucleus mass m. The main assumption in this case is that the motion of an electron can be considered at a fixed nucleus. As a result, it is possible to obtain a potential energy that depends exclusively on the coordinates of the nucleus, which allows its motion to be described separately [4].

The next rather general approximation is the assumption of the classical character of the motion of nuclei, which is violated at low temperatures and high densities. The condition of applicability is the restriction imposed on the ratio n of the thermal de Broglie wavelength of the nucleus $\Lambda = h(2\pi m k_B T)^{-1/2}$ to the radius of the volume per nucleus, $R_0 = [3m/(4\pi\rho)]^{1/3}$:

$$\eta = \frac{\Lambda}{R_0} = \left(\frac{2}{9\pi}\right)^{1/6} \frac{h\rho^{1/3}}{m^{5/6}(k_B T)^{1/2}} \ll 1.$$

Here ρ is the mass density, m is the mass of the nucleus, and T is the temperature.

Electrons in matter are conditionally divided into bound and free electrons. The latter are also called conduction electrons. Bound electrons are localized near nuclei due to a strong electron–nuclear interaction, move together with nuclei and always exhibit quantum properties. With increasing temperature and density, bound electrons become free. The role of quantum effects in a gas consisting of free electrons is determined by the ratio of the temperature ($k_B T$) to the Fermi energy $\varepsilon_F = (h^2/2m_e)(3n_e/8\pi)^{2/3}$, where n_e is the electron density.

Condensed matter is characterized by the inequality $k_B T \leqslant \varepsilon_F$. Under this condition, the quantum properties of free electrons become dominant. The theoretical description can be simplified provided that the thermal energy density does not reach a value at which the bound electrons can undergo a transition into an excited state or partially break away. At high temperatures, when $k_B T \gg \varepsilon_F$, free electrons behave like classical particles.

When the density decreases and the temperature rises, the condensed substance passes smoothly into the gas-like state or crosses two-phase regions. For metals (under normal conditions), the conditional boundary between gaseous and liquid states is the position of transition from metal to dielectric. In the plasma state, there is a region of a relatively weak interaction that is characterized by small values of the nonideality parameter $\Gamma = Ze^2/(4\pi\varepsilon_0 R_0 k_B T)$, representing a ratio of the average potential energy to the average kinetic energy.

The boundary between the plasma and the gas in the density–temperature plane can be determined by setting a small value of the degree of ionization of the atoms.

The gas phase of the substance corresponds to a system of neutral weakly interacting classical particles.

The authors of many papers neglect the effect of temperature and consider such a system as highly degenerate, formally setting $T = 0$. These authors calculate the equation of state of a cold substance (for example, helium) at high pressure and study in detail the transition to the metallic state.

The domain of dense plasma is least convenient for studying at intermediate temperatures. The most general method that can be applied to this domain is to use classical mechanics to describe the motion of nuclei. In this case, we consider a model that represents a plasma as a set of electrically neutral weakly interacting cells with a volume V_0. Strictly speaking, the domain of adequate applicability of this model is unknown, but the limits of its physical inconsistency can be determined using the concept of the electron radius of the Debye screening, $r_D = [\varepsilon_0 k_B T / (e^2 n_e)]^{1/2}$. If free electrons cannot screen an ion in a cell of volume V_0, then there are no physical grounds for using such a cell model. In a gas consisting of degenerate electrons, the screening length is defined as $r_{TF} = (\pi / 3 n_e)^{1/6} [h^2 \varepsilon / (4 \pi m_e e^2)]^{1/2}$ and can be employed to determine the applicability limit of the cell model. Let us now consider the Wigner–Seitz cell approximation. Each cell contains a nucleus with a charge Z and a corresponding number of electrons, which ensures electrical neutrality. The motion of electrons in the field of the nucleus inside this cell is described with the help of the chosen physical model. The effect of the electrons and nuclei from neighboring cells is taken into account through the boundary conditions. The contributions of nuclei and electrons are usually described independently in accordance with the adiabatic approximation. In practice, a real Wigner–Seitz cell of a complex geometric shape is usually replaced by a spherical cell of volume V_0.

Let us now say a few words about the physical models describing the electron subsystem. One of the standard approaches is the Green's function method [5]. Another widespread approach is presented by the Singwi–Tosi–Land–Sjölander theory [4].

The most consistent and effective methods include density functional theory (DFT) (see section 5.3). This method is based on the theorem that states that the free energy of an electronic system in an external field can be expressed as a functional of the electron density. This functional is usually unknown, and the use of any of the approximations gives rise to so-called quantum-statistical models or quantum-mechanical Hartree models. The employment of the DFT approach turned out to be very successful in condensed state physics.

The first statistical model was the Thomas–Fermi model, in which the electron density is calculated in the semiclassical approximation using a self-consistent potential [6]. When regular and oscillatory corrections were taken into account, the theoretical results started to reproduce the experimental ones and shell effects were included in the thermodynamic model.

The range of applicability of the Thomas–Fermi models, determined by comparison with the results of dynamic experiments and for condensed densities, corresponds to pressures $p \geqslant 10^2$ GPa.

An even more consistent description of the bound states in a plasma is realized using quantum-mechanical self-consistent field models in which the wave functions of electrons are solutions of the Schrödinger equation with the boundary conditions of the model. The latter represent the translational symmetry of the crystal lattice or an approximate description of this symmetry. The first formulations of quantum-mechanical models applied the Thomas–Fermi approximations in combination with electrostatic theory.

The Hartree–Fock model, which takes into account the exchange interaction of electrons, is the most complete one-electron model, but at the same time the most cumbersome. A simpler description of the exchange effects is attained in the Hartree–Fock–Slater approximation using a so-called local effective exchange potential. In concrete calculations based on the Hartree–Fock–Slater model, a quasi-classical approximation was used for strongly excited states of bound electrons and for free electrons. Another researcher [7] proposed very simple boundary conditions, allowing the upper and lower boundaries of the energy bands of electrons to be simulated. Sinko [8] closed this model by an approximation for the density of states having the shape characteristic of free electrons, and proposed the so-called self-consistent field model.

As was mentioned above, the adiabatic approximation allows one to separately calculate the contributions of electrons and nuclei to various thermodynamic functions. In earlier works, the nuclei were treated as an ideal gas. A model of point ions in a medium with a uniformly distributed negative charge was developed, which takes into account the nonideality of the motion of the nuclei.

The general drawback of the above-described cell models is the impossibility of accounting for interparticle correlations at distances exceeding the size of a unit cell. Cell models are not capable of describing typical plasma states characterized by long-range correlations when the Debye sphere contains a large number of particles. The bounded atom model combines ideas of both solid state and plasma; an even simpler approach is the chemical model of plasma described below.

5.2 Chemical model of plasma

5.2.1 Thermodynamics of shock-compressed plasma at megabar pressures: nonideality and degeneracy

The approach based on a quasi-chemical representation, i.e. the chemical model (conditionally called SAHA), has been used repeatedly to describe the thermodynamic properties of plasma in a wide range of parameters—from states close to those of condensed matter to a rarefied solar plasma [9]. Below we present the results of applying the SAHA approach for describing hydrogen (deuterium) and inert gases compressed by high-power shock waves.

Experiments on shock compression of gases cover the pressure range up to 100 GPa and higher and densities up to 0.8 g cm^{-3} for deuterium and 10 g cm^{-3} for xenon. The range of parameters studied in the experiments is characterized by extremely complex and diverse processes that must be reflected in the corresponding physical models. First of all, as the substance is compressed, the component composition of the medium can change sharply, which is accompanied by the appearance of a strong interparticle interaction: Coulomb interaction between electrons and ions, polarization interaction between charges and neutrals, and short-range interaction between neutral particles. Since the characteristic interparticle distance in the medium in question is comparable to the characteristic size of atoms and ions, part of the phase volume occupied by them becomes inaccessible to other particles, which leads to an increase in their kinetic energy and corresponding contributions to the free energy of such highly compressed disordered structures. In addition, strong compression causes a change in the energy spectrum of the bound states of atoms and molecules. With increasing compression, one should also take into account the change in the statistics of the electrons of the continuous spectrum from Boltzmann to Fermi statistics, since the degeneracy parameter $n_e \Lambda_e$ can increase several-fold under these conditions.

A version of the pseudopotential model for multiple ionization was used to describe the Coulomb interaction [10]. The central point of this model is the explicit allowance for the 'nonCoulomb' nature of the interaction of free charges at close distances, which under conditions of strong nonideality leads to a noticeable positive shift of not only the potential but also of the average kinetic energy of free charges. In this case, following [10], the depth of the electron–ion pseudopotential Φ_{ei}^* was related to the boundary separating the free states of each electron–ion pair and its bound states in the partition function.

Figure 5.1 compares this potential with other pseudopotentials proposed for plasma.

From figure 5.2 we can conclude that for equivalent pseudopotentials the results are in satisfactory agreement with the results of direct numerical simulation by the Monte Carlo method. This can be regarded as a confirmation of the conclusion drawn in [11] that the observance of general relations is a key condition for constructing a satisfactory description of the thermodynamics of strongly nonideal Coulomb systems.

For the case of partially ionized xenon plasma, calculations were made [12] using the above variant of the pseudopotential model [10]. In these calculations, the

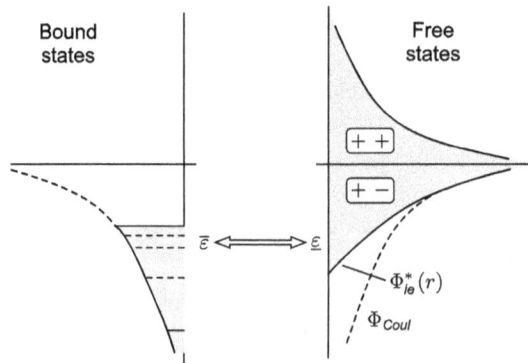

Figure 5.1. Electron–ion Glauberman pseudopotential: ε and $\bar{\varepsilon}$—a movable mutual boundary that limits the energies of free and bound states.

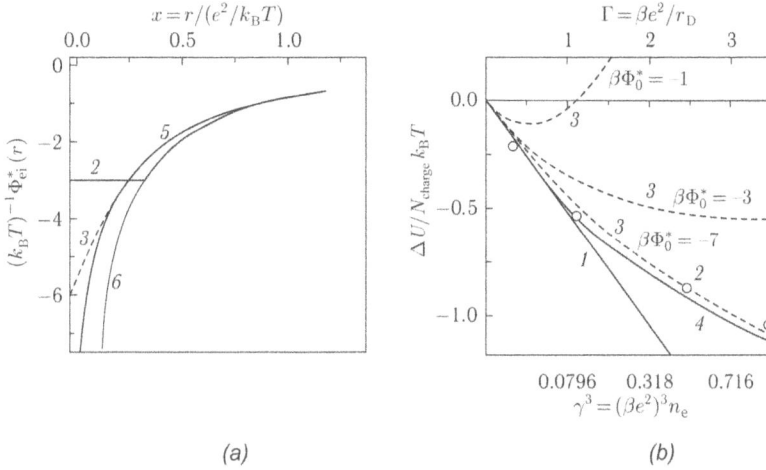

(a)   (b)

Figure 5.2. Effective electron–ion pseudopotential and dimensionless configuration energy: (a) the effective electron–ion pseudopotential $\Phi^*_{ei}(r)$ (divided by $k_B T$) [10]: 2—simplified ('zero model') electron–ion pseudopotential $\beta\Phi^*_{ei}(0) = -3$; 3—equivalent pseudopotential $\Phi^*_{ei}(r)$ from [10] ($\beta\Phi^*_{ei} = -6$); 5—electron–ion pseudopotential of hydrogen at $T = 10^3$ K; 6—Coulomb potential. (b) the dimensionless configuration energy $\Delta U/(N_{charge} k_B T)$ of the free charge subsystem: 1—Debye approximation (the Debye limit); 2—Norman configuration energy calculated by the Monte Carlo method for the simplified ('zero model') electron–ion pseudopotential with $\beta\Phi^*_{ei}(0) = -3$; 3—linearized version of the equivalent pseudopotential $\Phi^*_{ei}(6)$, $\beta\Phi^*_{ei}(0) = -7$; 4—nonlinear approximation for the equivalent pseudopotential $\Phi^*_{ei}(r)$ with $\beta\Phi^*_{ei}(0) = -7$.

boundary separating the free and bound states of an atom and an ion and the corresponding limitation of the atomic partition function were chosen at a binding energy depth of the order of $k_B T$, which virtually coincides with the well-known and often recommended procedure for calculating the partition function by the so-called Brillouin–Planck–Larkin formula (see more details in [3]). The comparison showed (figures 5.3 and 5.4 [12]) that the proposed variant of the pseudopotential model [10] allows qualitative and, taking into account the actual accuracy and scatter of the experimental data, quantitative description of the parameters of the experimentally measured shock adiabats of xenon.

Figure 5.3 compares the experimental data with the results of calculations by the model [12]. The shift of the shock adiabats obtained in the model [10] is directly related in this case to the presence of an explicit (positive) correction to the average kinetic energy of free charges in the model, which in terms of the equations of state is equivalent, in comparison with most traditional approximations, to the effect of additional repulsion.

A similar result was obtained in comparative calculations using the model [10] of parameters of a shock-compressed cesium plasma. In these experiments, a higher measurement accuracy was achieved, and a wider range of pressures and specific volumes and, what is even more important, a wider range of plasma nonideality parameters r_D (because of the extremely low cesium ionization potential) were involved. Another distinctive feature of the experiments and the accompanying series of computational and theoretical studies [16] is apparently one of the first

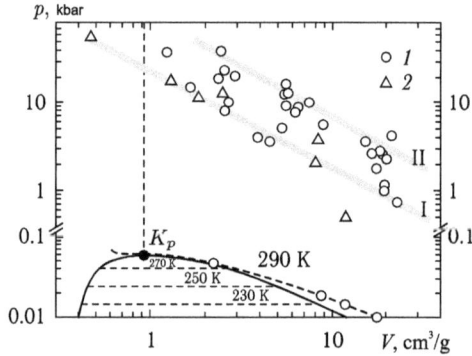

Figure 5.3. Phase diagram of xenon: marked are the boundaries of the two-phase region and the critical point K_p; 1—experimental data [12]; 2—experimental data [13] after correction [14] of the initial density of the unperturbed gas before the shock jump with allowance for the equation of state of a real gas; the dashed lines are isotherms in the two-phase region [14], the dash-dotted line is the critical isochore ($V = V_c$), the gray lines are the boundaries of single-ionization {$\chi_{Xe^+} = \chi_{Xe}$, ($\chi_{Xe} = N_{Xe}/N_{total}$)} and double ionization II ($\chi_{Xe^{++}} = \chi_{Xe^+}$).

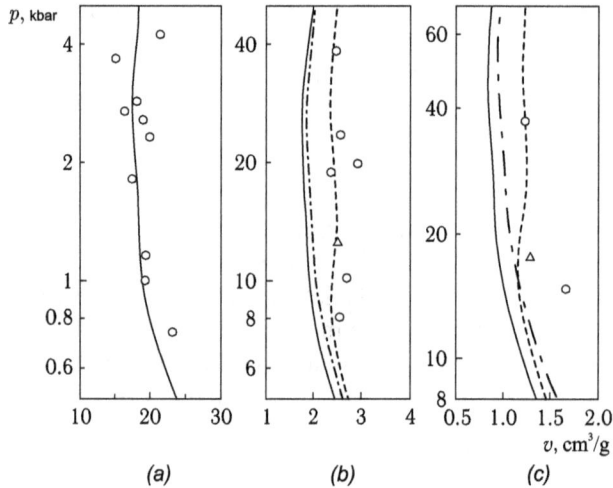

Figure 5.4. Phase diagram of xenon: comparison of experimental shock adiabats and results of model calculations; the solid curves show the interaction between charged particles taken into account in the modified ring Debye approximation in a grand canonical ensemble; the dash-dotted curve illustrates the atom–atom interaction added in the approximation of the second virial coefficient [15]; the dashed curve demonstrates a calculation based on a modified pseudopotential model [10], taking into account the equality of the depth of the pseudopotential of free particles and the upper cutoff boundary of intra-atomic states, chosen equal to k_BT. The pressure in the front of the shock wave: (a) $p_0 = 1$, (b) $p_0 = 10$ bar, and (c) $p_0 = 20$ bar.

(if not the first) realization of the well-known idea of Ya B Zel'dovich [17] on temperature recovery from the results of shock-wave experiments. However, the accuracy of the thus recovered temperature and the specific effect of mutual compensation of various uncertainty sources in the model equation of state of cesium for this range of parameters have led to the fact that the finally recalculated 'experimental' results for thermal $p(V, T)$ equation of state of cesium proved to be compatible with almost all theoretical approximations, subsequently proposed to describe these experimental data.

5.2.2 Gas thermodynamics

In selecting the parameters of the model [10] for shock-compressed initially liquefied inert gases, the atomic radii (for a fixed energy constant ε_{SS}) and the repulsion degree in the soft sphere mixture approximation were chosen proceeding from the condition of the best description of the calculational cold curve data ($T = 0$ K) for the experimental range densities. The ratios between the radii of atoms and ions of different multiplicities were determined from the calculation of their electronic structure in the bounded approximation by the Hartree–Fock method. Figure 5.5 shows the dependence of the energy of the xenon atom and ions on the radius of the atomic cell.

The ratio of the radii of the atom and ions was chosen on the line, where the energy shift during compression was compared with the ionization potential of the atom, i.e. on the line $\Delta E/I = 1$.

Using the model [10], the shock adiabats and isotherms of xenon, krypton, and argon were calculated. In all three cases, the initial state of the shock-compressed gases corresponded to a liquid. Figure 5.6 shows the shock adiabat of xenon in the density–pressure coordinates.

Along with shock experiments, the data on multiple shock compression are also presented [22]. One can see that, in general, using the adopted approximation, it is possible to achieve a satisfactory description of the experimental data. The discrepancies at low pressures and temperatures can be explained by insufficient accuracy of the approximation of states of liquid-phase xenon.

The model also permits shock adiabats of liquid argon and krypton to be satisfactorily described (figures 5.7 and 5.8). The experimental data were borrowed from papers [24, 25].

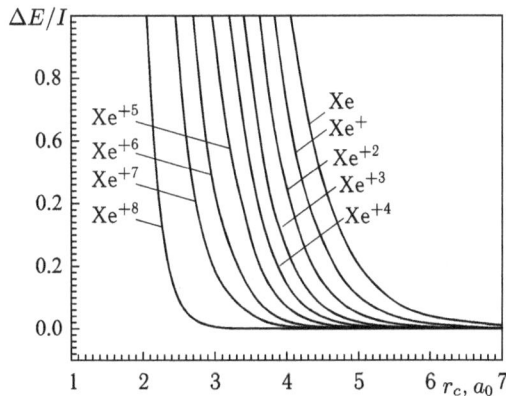

Figure 5.5. Energy variation of atom and ions of xenon compressed in a spherical cell as a function of ionization potential of an isolated atom.

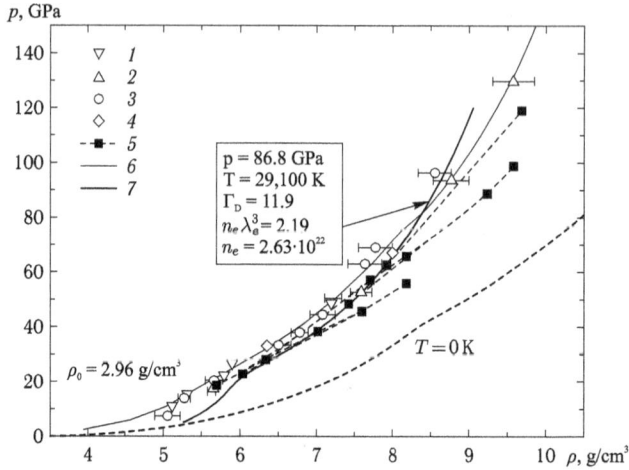

Figure 5.6. Shock adiabat of xenon: experimental data; 1—[18], 2—[19], 3—[20], 4—[21], 5—[22]. Calculated curves; 6—[20], 7—[23], the dashed curve is the 'cold curve' [20].

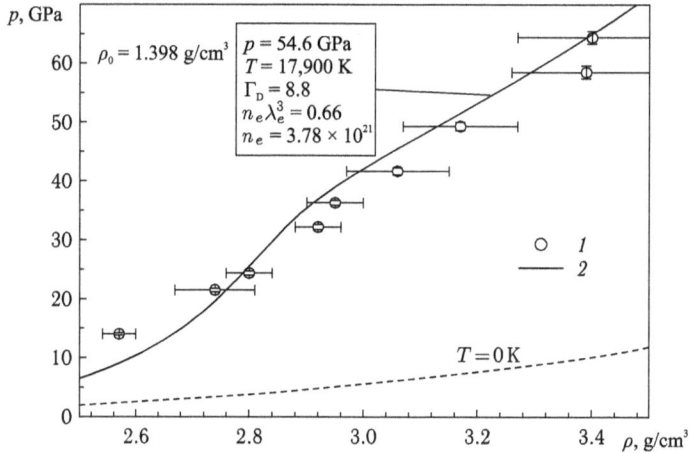

Figure 5.7. Shock adiabat of argon. Experimental data; 1—[24]. The calculated curve; 2—present work; the dashed curve is the cold curve.

Note that a satisfactory agreement can be reached with the measured values of brightness temperature and speed of sound in these substances.

The above comparison of the calculation results with the experimental data on shock-compressed inert gases shows that the model [10] qualitatively and, in most cases, quantitatively describes the behavior of nonideal plasma in megabar and sub-megabar pressure ranges of shock compression of inert gases.

Note that with the general trend being preserved in the behavior of thermodynamic quantities, there exist nevertheless regions where the calculated data clearly deviate from the experiment. In particular, this can be observed due to a simplified

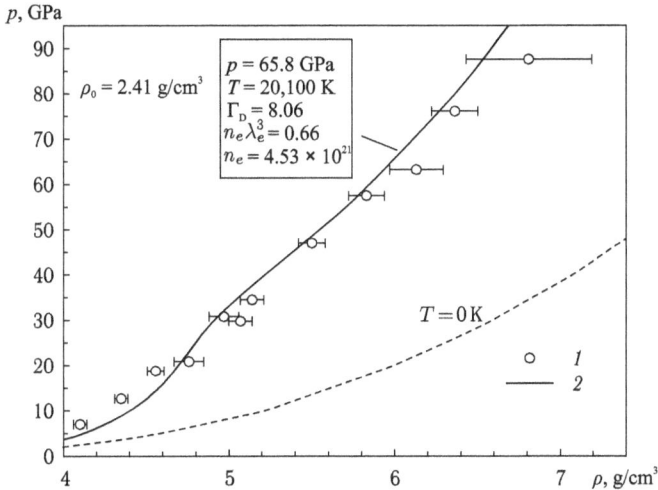

Figure 5.8. Shock adiabat of krypton. Experimental data; 1—[25]. The calculated curve; 2—[10]; the dashed curve is the cold curve.

definition of the repulsion parameters in the soft sphere model, which may require further improvements in this procedure.

5.2.3 Metal plasma thermodynamics

To date, the bulk of experimental information on the properties of highly compressed plasma has been obtained by dynamic methods [26], using the technique of high-power shock waves for compression and irreversible heating of substances. The use of explosive and pneumatic propellant devices in such experiments has made it possible to study and analyze theoretical models of thermodynamic, electrophysical, and optical properties of shock-compressed cesium, inert gases, and hydrogen under conditions of strong nonideality. The ionization multiplicity ($\alpha = n_e/(n_\alpha + n_i)$) of such a medium did not exceed one or two. Going beyond these conditions—the transition to the parameters at which a substance becomes multiply ionized with partially degenerate electrons—can be accomplished by drawing experimental data on the compression of solid and porous metals by shock waves with an amplitude pressure of hundreds of thousands or millions of atmospheres. To date, a significant amount of experimental data on the dynamic compression of metals has been obtained (see [27–29] and references therein) using shock waves generated by chemical [27, 30] and nuclear [31] explosions, by pneumatic propellant devices, and recently, by concentrated laser [32], x-ray, and ion fluxes. Data on shock-wave compression, supplemented by the results of recording the adiabats of unloading shock-compressed metals, form the basis for constructing semiempirical equations of state by choosing the optimal constants in functional thermodynamic relationships based on simplified thermodynamic models. At the same time, in the process of shock compression, melting occurs already at relatively low (100–200 GPa) pressures, followed by progressive thermal ionization and pressure-induced ionization of the substance.

Thus, a dense, disordered, multiply ionized system of charged particles, i.e. an electron–ion medium with a complex spectrum of intense collective interactions, is realized. For this reason, shock-compressed metals seem to be an interesting object for testing theoretical models of highly compressed plasma both for searching for plasma phase transitions and for analyzing various models describing the non-ideality of a highly compressed plasma at high energy densities. In essence, we speak of the extension of plasma models [33] to an unconventional region of condensed densities and megabar pressures, where either semi-empirical approximation equations of state [33] or far extrapolations of quasi-classical approximations have been used until recently. The presence of such thermodynamic measurements in the metal–dielectric transition region would also allow a hypothesis [34] to be verified, concerning the relation between metallization and the first-order phase transition in disordered media.

The range of parameters of a strongly nonideal system in question corresponds to lower densities ρ_0 (than those in solids) and to energies exceeding the binding energy of atoms and molecules in a solid (about 1 eV per particle). To generate such states of metals, use was made of shock-wave compression of finely dispersed (porous) metals, which makes it possible to increase the effect of irreversible energy dissipation at the shock front and to provide higher substance heating. For some metals, the porosity value $m = \rho_0/\rho_{00}$ (ρ_{00} is the density of the porous sample) lies in the range $1 \leqslant m < 30$, so that the experimental data covered a considerable region both in substance densities behind the shock front and in temperatures. For nickel, the maximum possible porosity m is equal to 15, 20, and 28, and the shock compression pressure is higher than 80 GPa [35, 36]; for copper, $m = 10$ [37]; for iron, $m = 20$ [38]; and for aluminum, $m = 8$ [39].

To point out the general features of the above experimental data on the shock compression of porous metals, we consider, following [38], the internal energy–density (E–ρ) diagram, supplemented by the calculation of isotherms of matter using the plasma model of a bounded atom. Figures 5.9 and 5.10 show an E–ρ diagram and a diagram of equilibrium composition for nickel.

Figure 5.9 demonstrates the fact that the entire phase diagram of the substance (the diagrams of other metals are similar to that under discussion) breaks up into two regions of qualitatively different behavior of thermodynamic dependences.

The most part is occupied by a relatively rarefied ($\rho \ll \rho_0$) gas plasma region, characterized by two distinctive features—a smooth decrease in energy under isothermal compression and a clear manifestation (at large intervals of density variation) of the so-called shell oscillations of all thermodynamic dependences (see details in [41]). For $\rho > \rho_0$, this behavior is replaced by a sharp increase in the energy and the generalized compressibility factor $Z(n_{\text{nucl}}, n_{\text{e}}, T) = p/p^{id}$, which is tradition-ally interpreted as pressure-induced ionization. This process in the limit of a very high density is completed by reaching the region of states that are well described by the model of a system of mobile nuclei immersed in a weakly nonideal gas of degenerate electrons. Thermodynamics in this region is successfully described by a well-developed apparatus of cell representations [6, 8, 42]. Between the two above-mentioned regions of the parameters, there is a transition region characterized by

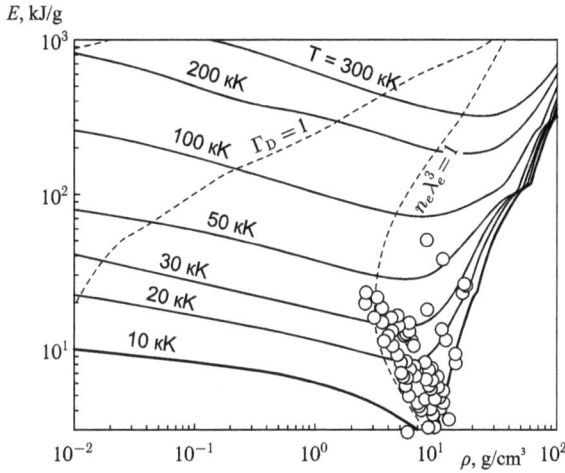

Figure 5.9. Density–energy diagram for nickel plasma. Plotted are the calculated isotherms and the experimental points obtained by shock compression of solid and porous nickel samples [37]. Marked are the lines of constant parameters of the Coulomb nonideality ($\Gamma_D = [4\pi(e^2/k_BT)^3 n_\alpha z^2_\alpha]^{1/2}$) and the electron degeneracy parameter $n_e\lambda^3_e$.

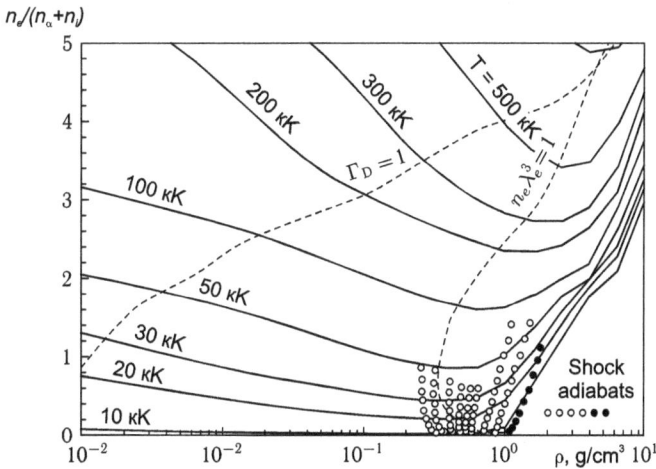

Figure 5.10. Behavior of the degree of nickel plasma ionization under compression [40]. The degree of ionization along the isotherms (solid curves) and calculated shock adiabats (points) of solid and porous nickel samples corresponding to the experiment are shown [36]. Marked are the lines of constant parameters of the Coulomb nonideality ($\Gamma_D = [4\pi(e^2/k_BT)\sum n_\alpha z^2_\alpha]^{1/2}$) and the electron degeneracy parameter ($n\lambda^3_e = 1$) in plasma.

minimum internal energy and compressibility factor and by maximum violation of conditions of weak nonideality. The depth and location of the minima, corresponding to the thermal and caloric equation of state, on the isotherms can be conditionally considered as the focus of maximum uncertainty of our knowledge on the thermodynamic properties of a compressed and heated substance. Note that the region where the minima are located corresponds to the so-called valley of nonideality. The uniqueness of shock-wave compression of porous targets is that it gives information about the behavior of a dense strongly nonideal medium in this most complex and interesting region.

The description of the states of shock-compressed porous metals was based on a quasi-chemical approach (chemical model) [10, 11]. The results of calculations of shock adiabats for porous nickel, copper, aluminum, and iron using the model [10] are presented in the following figures.

Figure 5.11 shows a comparison of the calculated and experimental data for the shock adiabats of iron. The data on iron indicates that the completely satisfactory agreement between the results of theoretical calculations and experimental data at high porosities (and maximum attainable degrees of expansion of an initially condensed metal) gradually deteriorates as we pass to the region of an increasingly dense plasma in experiments with compression of low-porous samples. Note that repulsion (a combination of intrinsic particle sizes) and attraction parameters employed in the calculation model for describing the new experimental data were chosen by using the same scheme as in the previously performed calculations [44].

Similar calculations were made for porous copper. Figures 5.12 and 5.13 show a comparison of the calculated and experimental data for the shock-compressed copper and nickel. Note that in figure 5.12, as in previous cases, the previously obtained results [44] were supplemented by new experimental data [37, 39].

On the whole, a comparison of the calculated and experimental data for nickel, iron, and copper shows that using even an extremely simplified approximation, a quasi-chemical representation allows one to satisfactorily describe the experimental data on shock compression of samples of sufficiently high porosity.

A good coincidence of new data with the results of previous calculations serves as additional evidence of the correctness of the initial assumptions embedded in the calculation procedure relying on a generalized chemical model with the aim of expanding its extrapolation capabilities. It should be emphasized that the additional

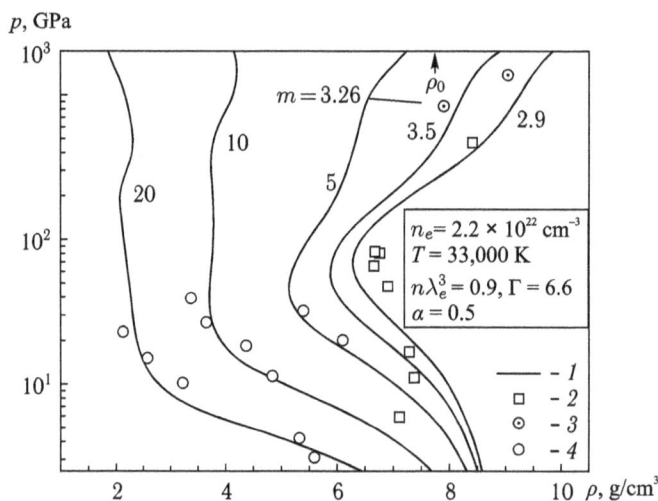

Figure 5.11. Shock adiabats for porous iron. 1—calculation by [10]; 2—experimental data [35] ($m = 2.9$), 3—experimental data [35]; 4—[43].

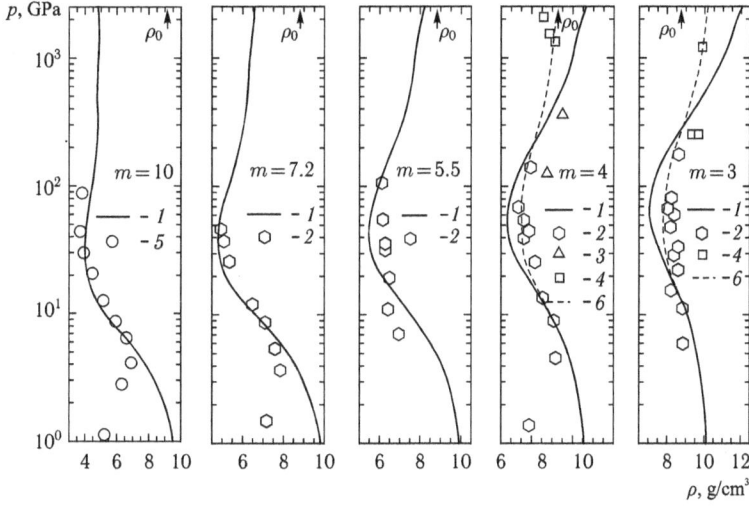

Figure 5.12. Shock adiabats for porous copper. 1—calculation by [10]; 2—experiment [35]; 3—experiment [45]; 4—experiment [46]; 5—experiment [37]; 6—calculation [47] by the model [10] with varied radii of the copper atom and ions [$r_\alpha = 2.0_{a0}$, r_c (Cu^{+1}–Cu^{+3} = 1.75a_0)].

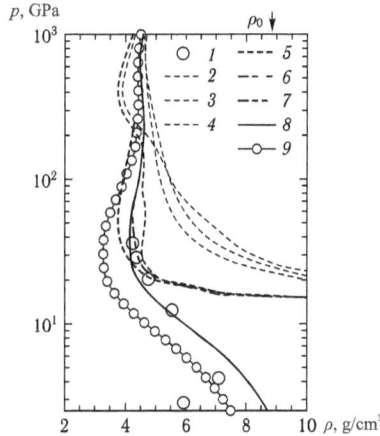

Figure 5.13. Shock adiabats of porous nickel for $m = \rho_0/\rho_{00} = 10$ [40]. Comparison of the results of calculations using different approximations: 1—experimental data [35, 36]; 2—approximation of an ideal plasma (taking into account only the ground states in the calculation of partition functions); 3—same as 2, but taking into account the Coulomb interaction; 4—same as 3, but the partition functions of atoms and ions are calculated by the Planck–Larkin partition function; 5, 6, and 7—same as 2, 3, and 4, but taking into account the short-range repulsion in the hard-sphere approximation; 8—same as 6, but with additional attraction with $e = 1$ (only for atoms); 9—same as 8, but with increased (+20%) radii of atoms and ions.

experimental data under discussion correspond not only to maximum (for a given range of nickel porosity) pressures ($p \approx 50$ GPa), but also, according to the present calculations, to maximum experimentally attainable temperatures and ionization degrees (for a shock-compressed nickel plasma). At the same time, as already mentioned above, the extrapolation capabilities of the chemical model gradually deteriorate at the same pressures in passing to the adiabats characterizing a lower porosity and correspondingly higher densities.

Thus, we can state a satisfactory, in general, correspondence between theory and experiment. At the same time, noteworthy is the discrepancy between the calculated and experimental data for the case of the adiabat with $m = 10$ (see figure 5.12) in the upper part of the experimentally reached pressure range. Analyzing the reason for this discrepancy, one should take into account the obvious simplicity of the theoretical model describing the interaction in the system and the extreme sensitivity of the results obtained to the specific choice of a combination of the particle's intrinsic sizes.

Note that the simplified scheme for choosing the relative size of the atom and ions, which gives satisfactory results for the description of shock-compressed porous iron, nickel, and copper, does not lead to any acceptable results in describing the shock adiabats of porous aluminum. Therefore, the above-described simplified procedure for determining the ratio of atomic and ionic radii has been modified. As the initial data for calculating this ratio, use was made of the results of calculating their electronic structure by the Hartree–Fock method.

5.3 Density functional method

The expression for energy in the Thomas–Fermi method at $T = 0$ can be written in the form of an explicit density functional [6]:

$$E\{n\} = T[n] + E_e[n] + E_i[n] \tag{5.1}$$

$$T[n] \sim \int dx \, n^{5/3}(x). \tag{5.2}$$

The minimum of the functional $E - \mu \int dx \, n$ in n leads to the Thomas–Fermi equation and determines the energy of the ground state of the system and the corresponding density distribution. Going beyond the framework of the Thomas–Fermi method and taking into account the corresponding additions, we easily see that we are still dealing with functional (5.1); however, functional (5.2) has a more complicated form. In particular, the account for quantum effects leads to the fact that instead of a quasi-uniform functional we obtain the functional that depends on the derivatives of n. It is not surprising, therefore, that the following assertion holds: when the many-body problem (many-particle Schrödinger equation) is precisely formulated, the energy of the system is expressed as functional (5.1) with a single-valued universal functional E_k of general form, the minimum of the functional $E - \mu \int dx \, n$ yielding energy and density distribution in the ground state of the system. This assertion was substantiated by March and Murray [38] within the framework of perturbation theory and proved as a rigorous theorem by Hohenberg and Kohn [49].

The density functional method (DFM) is an exact quantum-mechanical theory for a system of interacting quantum particles in an external potential $V_{ext}(r)$. The method itself is based on two strictly proved theorems [49].

1. For any system of interacting particles in the external potential $V_{ext}(r)$, the potential $V_{ext}(r)$ is determined up to an arbitrary constant by the electron density $n_0(r)$ in the ground state.
2. The energy of the nondegenerate ground state of the system for any external potential $V_{ext}(r)$ is a functional of the electron density $E[n(r)]$. The ground state of the system is the minimum of this functional, which is reached at a density $n_0(r)$ corresponding to the ground state of the system.

It follows from the first theorem that the Hamiltonian of the system is determined up to a constant value by the electron density $n_0(r)$ of the ground state; hence, it implies that many-particle wave functions are defined for all (ground and excited) states. Thus, all the properties of the system are completely defined if the electron density $n_0(r)$ of the ground state is known. The second theorem implies that for a known functional $E[n]$, we can find the density and energy of the ground state.

The energy functional in the formulation of [49] can be written as:

$$E_{HK}[n] = T[n] + E_{int}[n] + \int d^3r \ V_{ext}(r)n(r) + E_{II}$$
$$\equiv F_{HK}[n] + \int d^3r \ V_{ext}(r)n(r) + E_{II} \tag{5.3}$$

where E_{II} is the interaction energy of the nuclei. The functional $F_{HK}[n]$ includes the kinetic and potential energy of a system of interacting electrons:

$$F_{HK}[n] = T[n] + E_{int}[n]. \tag{5.4}$$

The minimum of functional (5.3) determines the energy of the system in the ground state and its electron density. Note that (5.3) does not carry any information about the excited states of the system.

The density functional method has a rigorous theoretical justification, but establishing a relationship between the electron density of the ground state and the properties of a substance is a nontrivial task. The main problem here is the fact that the kinetic energy is not directly related to the electron density function in the general case. To solve this problem, Kohn and Sham [50] suggested that the ground state of the system of interacting particles coincides with the ground state of an equivalent system of noninteracting particles, and the interaction can be taken into account by using the so-called exchange–correlation functional, which depends on the electron density. The problem in this formulation can be solved numerically up to an accuracy determined by the expression for the exchange–correlation functional. This technique proved to be very successful, and all existing calculations based on the density functional method utilize this approximation. The recently developed exchange–correlation functionals describe well the semiconductors of groups II–V, simple and transition metals, and insulators, for example, diamond, NaCl, and molecules with covalent or ionic bonds. Thus, the Kohn–Sham approach [50] is a very important step in the development of methods for calculating various properties of many-electron strongly interacting systems.

The Kohn–Sham equations for finding the ground state of the system are solved in different ways. The most widespread ones are listed below.

The most natural basis for the expansion of wave functions in the Kohn–Sham equations is a plane-wave basis. Nearly free electrons are well described by this basis; in addition, plane waves are very convenient for calculating the band structure of matter. Various pseudopotentials, both empirical and theoretical, are used to describe the electrons of inner shells [51, 52]. The method of plane waves has found wide application in quantum molecular dynamics calculations [53].

Another approach consists in constructing the wave functions of a many-electron system by combining the wave functions of individual atoms, the so-called localized orbital method or the strong-coupling method. There are many versions of this approach, the most well-known being the method of linear combination of atomic orbitals [54]. The strong-coupling method is the simplest and fastest technique for calculating the band structure of a substance; it forms the basis of numerical approaches, for which the simulation time depends linearly on the number of particles. This method is also the basis of more complex methods, for example, a linear combination of muffin-tine orbitals [55].

Finally, there is an approach combining the advantages of the two previous ones, the so-called method of adjoint functions [56]. For the inner shells of an atom, the method of localized orbitals is used, with the spherically symmetric Kohn–Sham, Schrödinger or Dirac equations being normally solved. For a region of space between atoms, the solution is found through expansion in basis functions (plane waves). This computational method is most often used in so-called all-electron calculations when all the electrons of an atom are taken into account; on the other hand, this method is the most time-consuming.

The thermodynamic functions are calculated using the density functional method as follows. The calculation of the total energy of the electron system is the simplest; at $T = 0$, one can also calculate the pressure, $p = -dE/dV$. At nonzero temperature, the self-consistent calculation uses the occupation numbers proportional to the Fermi–Dirac functions:

$$f(\varepsilon, \rho, T) = \frac{1}{1 + \exp[(\varepsilon - \mu(\rho, T))/T]}, \tag{5.5}$$

where μ is the chemical potential of the system, determined from the electro-neutrality condition. By knowing the occupation numbers, one can calculate the configuration entropy [42] and the free energy of the system and determine all the necessary thermodynamic parameters.

The striking efficiency of the electron density functional method has opened up wide possibilities of its application for solving chemical, physical, biological, and nuclear-physical problems.

Their detailed description can be found in monographs [57–59] and also in a collection of lectures [60].

Following [61], we list only some of the directions of development of this method:
- spin-polarized systems—DFM with allowance for spin;
- systems with degenerate ground states;
- multicomponent systems (electron–hole droplets of the nucleus);
- statistical ensembles for degenerate ground states;
- free energy at finite temperatures;
- quasi-equilibrium ensembles for excited states;
- relativistic electrons, astrophysics;
- current-dependent functionals;
- time-dependent phenomena, excited states;
- bosons (instead of fermions);
- combination of density functional theory with the molecular dynamics method or Monte Carlo method (especially useful for determining geometric structures). Car–Parrinello method;
- combination of local density approximation with the Hubbard parameter; and
- properties of compressed nuclear matter and many other scientific directions.

We will mention only a few examples of DFM application, which are close to the subject of our monograph [62].

5.3.1 Atomic and molecular structures

Table 5.1 [62] presents experimental data on the total energy of light atoms and the deviations from the total-energy experiment calculated in the Hartree–Fock approximation (HF) ΔE_{HF}, as well as within the framework of the local spin density functional without ΔE_{LSD} and taking into account $\Delta E_{LSD\text{-}SIC}$ correction for residual self-interaction. The data is presented for two variants of correction: depending only on the orbital densities $\Delta E_{LSD\text{-}SIC}^{(1)}$ and depending also on the angular momentum $\Delta E_{LSD\text{-}SIC}^{(2)}$.

One can see from table 5.1 that the values of the total energies of the atoms in the HF approximation are higher than the experimental values. In the LSD approximation, they are even larger, with an error approximately twice that of the HF approximation. Allowance for the residual self-interaction correction leads to energies below the experimental ones, but in most cases lying closer to them than the energies in the HF approximation. Taking into account the explicit dependence of the residual self-interaction correction on the orbital momentum leads to a remarkable agreement with experiment for the total energies of the atoms in question.

The deviations of the calculated values of the first ionization potentials of light atoms from the experimental ones as a function of Z are shown in figure 5.14 [62]. The data obtained in the HF approximation are joined by lines: the dashed line stands for the case when the ionization potentials were estimated by Koopmans' theorem as the eigenvalues taken with the opposite sign, and the solid line is plotted by using the difference of the total energies of the neutral atom and ion. The circles correspond to the errors in the first ionization potentials obtained in the residual

Table 5.1. Deviations of the theoretically obtained total energies of light atoms from the sum of the experimental ionization potentials corrected for the effects neglected in the calculations (relativism, etc) (for the notation, see [62]).

Z	Atom	E_{exp}	ΔE_{HF}	ΔE_{LSD}	$\Delta E^{(1)}_{LSD-SIC}$	$\Delta E^{(2)}_{LSD-SIC}$
1	2	3	4	5	6	7
2	He	−79.0	+1.1	+1.9	0.4	−0.4
3	Li	−203.5	+1.3	+3.7	−0.7	−0.7
4	Be	−399.1	+2.6	+6.1	−0.7	−0.7
5	B	−670.8	+3.4	+8.3	−1.2	−0.7
6	C	−1029.7	+4.2	—	−2.3	−0.9
7	N	−1485.3	+5.0	+12.6	−3.6	−1.1
8	O	−2042.5	+6.9	—	−5.1	−1.2
9	F	−2713.5	+8.6	+16.9	−7.2	−1.7
10	Ne	−3508.1	+10.3	+19.2	−9.5	−2.1
11	Na	−4414.7	+10.5	+22.1	−11.4	−2.7
12	Mg	−5443.1	+11.6	+24.8	−13.3	−3.3
13	Al	−6594.0	+12.5	+28.0	−14.8	−3.2
14	Si	−7873.2	+13.5	—	−16.3	−3.2
15	P	−9285.1	+14.1	+34.0	−18.6	−3.3
16	S	−10 832.3	+16.2	—	−19.7	−3.0
17	Cl	−12 520.7	+18.2	—	−21.5	−2.6
18	Ar	−14 354.6	+19.9	+44.1	−23.7	−2.3

Note. The energies are expressed in electronvolts.
In the table, 1 at unit = 27.21 eV.

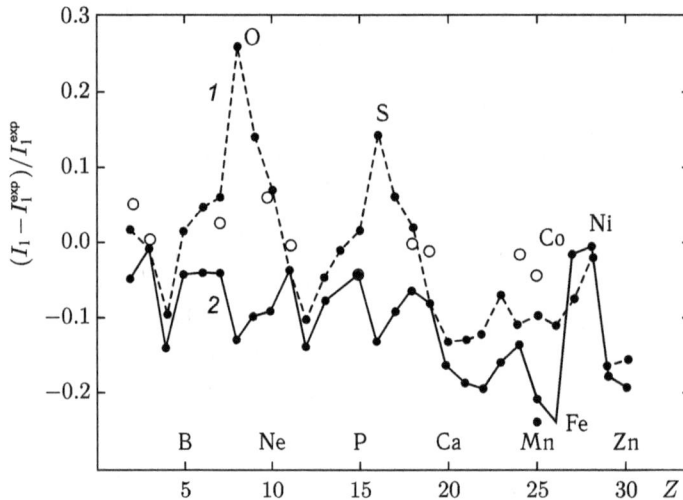

Figure 5.14. Deviation of the theoretical first ionization potentials of light atoms from the experiment. Curve 1 shows the HF approximation by Koopmans' theorem; curve 2 is the HF approximation with relaxation; the circles demonstrate the results of the LSD approximation.

Table 5.2. Experimental ionization potentials of light atoms and the results of their calculation by various methods [62].

Z	Atom	Experiment	HF method		LSD-SIC
			$-\varepsilon_{max}$	$E^{(+)}_{HF} - E^{(0)}_{HF}$	$-\varepsilon_{max}$
1	2	3	4	5	6
2	He	24.587	24.98	23.45	25.8
3	Li	5.392	5.341	5.341	5.4
4	Be	9.322	8.416	8.043	—
5	B	8.298	8.432	7.932	—
6	C	11.260	11.79	10.79	—
7	N	14.543	15.44	13.96	14.9
8	O	13.618	17.19	11.89	—
9	F	14.422	19.86	15.72	—
10	Ne	21.564	23.14	19.84	22.9
11	Na	5.139	4.955	4.952	5.1
12	Mg	7.646	6.884	6.615	—
13	Al	5.986	5.714	5.507	—
14	Si	8.151	8.081	7.660	—
15	P	10.486	10.66	10.04	10.0
16	S	10.360	11.90	9.031	—
17	Cl	12.967	13.78	11.80	—
18	Ar	15.759	16.08	14.78	15.8
19	K	4.341	4.011	4.005	4.3 —
20	Ca	6.113	5.320	5.118	—
21	Sc	6.54	5.717	5.347	—
22	Ti	6.82	6.008	5.513	—
23	V	6.74	6.275	5.668	—
24	Cr	6.766	6.038	5.88	6.7
25	Mn	7.435	6.743	5.90	7.1

Note: The ionization potentials are expressed in electronvolts; E^{0}_{HF} and E^{+}_{HF} are the energies of the neutral atom and ion in the HF approximation; ε_{max} is the maximum eigenvalue of the energy of occupied electronic states.

Table 5.3. Binding energy of an electron in negative ions [62].

Ion	HF	LSD-SIC	Experiment	Ion	HF	LSD-SIC	Experiment
H^-	−0.33	0.7	0.75	Cl^-	2.58	3.8	3.61
O^-	−0.54	1.6	1.46	Br^-	2.58		3.36
F^-	1.36	3.6	3.45	I^-	2.47		3.06

self-interaction corrected LSD approximation. The presented results show that the residual self-interaction corrected LSD approximation yields the first ionization potentials that are in better agreement with the experiment. The data used to plot figure 5.14 are listed in table 5.2.

Table 5.3 contains data on the binding energy of negative ions.

Table 5.4. Some characteristics of diatomic molecules, obtained in the local spin density approximation, and their comparison with the experiment [62].

Molecule 1	D, eV Experiment 2	D, eV LSD 3	R, at units Experiment 4	R, at units LSD 5	ω, cm^{-1} Experiment 6	ω, cm^{-1} LSD 7
H_1	4.8	4.9	1.40	1.45	4400	4190
Li_2	1.1	1.0	5.05	5.12	350	330
B_2	3.0	3.9	3.00	3.03	1050	1030
C_2	6.3	7.3	2.35	2.35	1860	1880
N_2	9.9	11.6(9.94)	2.07	2.07(2.07)	2360	2380
O_2	5.2	7.6	2.28	2.27	1580	1620
F_2	1.7	3.4	2.68	2.61	890	1060
Na_2	0.8	0.9	5.82	5.67	160	160
Al_2	1.8	2.0	4.66	4.64	350	350
Si_2	3.1	4.0	4.24	4.29	510	490
P_2	5.1	6.2	3.58	3.57	780	780
S_2	4.4	5.9	3.57	3.57	730	720
Cl_2	2.5	3.6	3.76	3.74	560	570

Quantum-chemical methods for calculating the characteristics of molecules, based on Slater determinants, lead to a sharp increase in the amount of computations with increasing number of atoms in the molecule and/or charge of the nuclei of its constituent atoms. Despite great difficulties in the calculations based on Slater determinants, the results obtained agree with the experiment not so well as those of much less cumbersome calculations based on the local spin density approximation.

Table 5.4 [62] lists the dissociation energies, the equilibrium interatomic distances, and the oscillation frequencies of diatomic molecules consisting of some light atoms, calculated within the local spin density approximation. The corresponding experimental data is also presented in the table.

The equilibrium distances and vibrational frequencies obtained within the framework of the LSD approximation well agree with those obtained experimentally: for the molecules included in table 5.4, the average deviation of equilibrium distances and vibrational frequencies from the experiment is 0.05 at units and 80 cm^{-1}, respectively. The dissociation energies are somewhat overestimated; the average deviation from the experiment, according to table 5.4, is 1.2 eV. The calculation for the nitrogen molecule N_2 shows (the results are given in parentheses in the corresponding row of the table) that the residual self-interaction corrected LSD approximation should substantially improve the agreement of the calculated dissociation energies with experiment.

5.3.2 Condensed media

The deviations between the calculated and experimental binding energies of atoms in the crystal and densities corresponding to zero temperature and pressure as functions of Z are shown for some metals in figures 5.15 and 5.16 [62].

The presented results were obtained, as noted above [62], by far not the most perfect method for calculating the band structure, and therefore should be considered only as an upper estimate of the accuracy that can be achieved at present in calculating the properties of crystals by the density functional method.

The calculations performed showed that, except for lanthanides and light actinides, the theory basically correctly predicts a stable structure. Only for sodium and gold, the bcc structure proves to be more advantageous than the experimentally observed hexagonal close-packed and face-centered cubic structures, respectively. Less successful was the prediction of stable structures of lanthanides and light actinides, although the number of coincidences with the experiment was also significant.

Apparently, this is due to the inadequacy of the calculation method for metals with an unfilled f-shell.

A known problem in the calculations of the band structure is the discrepancy between the experimental values of the energy gap band between the filled and empty bands in insulators and the results of calculations in the HF and local spin density functional approximations. If the HF approximation overestimates the

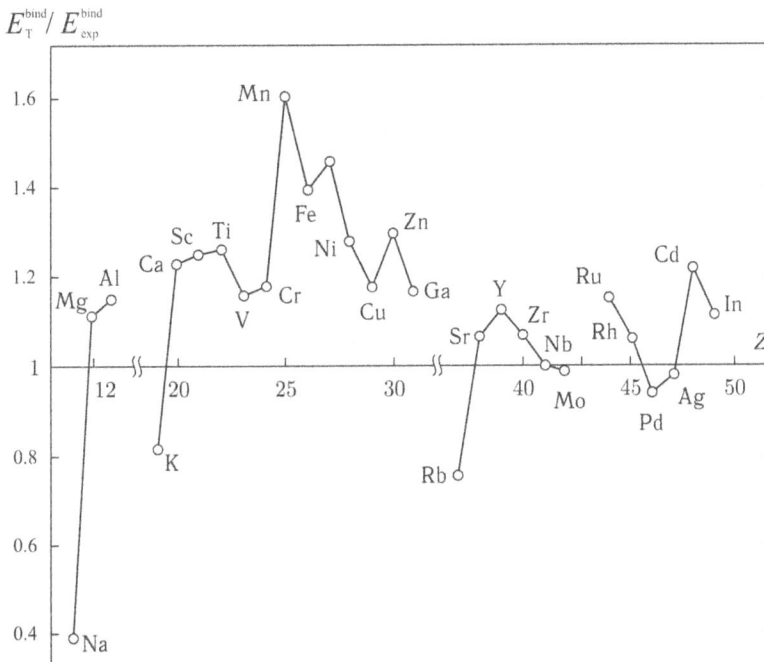

Figure 5.15. Deviations between theoretical and experimental binding energies of atoms in some metals.

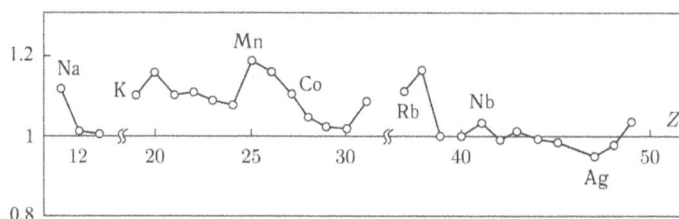

Figure 5.16. Deviations between theoretical and experimental crystalline densities of some metals at zero pressure and temperature.

Table 5.5. Energy band structure of noble metal crystals.

Crystal	ΔE_{exp}	$\Delta E_{HF} - \Delta E_{exp}$		$\Delta E_{LSD} - \Delta E_{exp}$	Δ_{SIC}
		without correlation	with correlation		
Ne	21.4	+3.8	+0.9	−10.2	+9.9
Ar	14.2	+4.3	+1.0	−5.9(−0.7)	+5.8
Kr	11.6	+4.8	+1.8	−4.9	+4.9

bandwidth due to the neglect of correlations, the reason for its underestimation in the LSD approximation can be due to the fact that because of the different degree of localization of the orbitals in the valence and conduction bands, residual self-interaction variously affects the corresponding eigenvalues. For example, p-like valence bands in noble gas crystals are narrow and separated from other bands, like atomic p-states. Therefore, we can expect that when the residual self-interaction correction is taken into account, the same shift of p-states, as that observed in atomic calculations, will take place in crystals. Table 5.5 [62] lists the experimental values of the band gap width ΔE_{exp} in noble gas crystals and their deviations from the results of calculations in the HF approximations without and with approximate allowance for correlations and the local spin density functional. Also presented are the differences ($\Delta_{SIC} = E^{LSD}_{appr} - E^{LSD-SIC}_{appr}$) between the energies of the external p-orbitals calculated without and with allowance for the residual self-interaction correction. All the data are given in electronvolts.

The data presented show that the displacement of the eigenvalues of the external p-orbitals in noble gas atoms, when taking into account the residual self-interaction correction, virtually coincides with the LSD approximation band structure error. This suggests that the difficulties of approximating the local density in this case are related to the residual self-interaction. The result obtained in the calculation for an argon crystal with allowance for the residual self-interaction correction (presented in parentheses in the corresponding row of the table) confirms this assumption, thus indicating that one should not *a priori* ignore the residual self-interaction in crystals.

The density functional methods are successfully used to calculate the thermodynamic properties of substances. The density functional method is most often used to calculate cold curves, isotherms $T = 300$ K, and shock adiabats of the crystalline phase [63–66]. For example, the room-temperature isotherms and shock adiabats of Al, Cu, Ta, Mo, and W are given in [63]; the cold curve of Zn, in [64]; the cold curve of Al and the room-temperature isotherm of Be, in [64]; the isotherm $T = 300$ K of Mg, in [65]; and the cold curve of Al up to ultrahigh pressures of about 100 Mbar, in [66].

We see that the density functional method, being equivalent to the exact solution of the many-body problem, is advantageous over direct quantum-mechanical methods in terms of simplicity and is therefore considered a powerful tool for describing a wide range of physical properties of complex electronic systems. In condensed state physics, it is sometimes called the standard model, emphasizing the high quality of its results.

This method does not replace, but complements for the previous approaches, revealing yet another way of constructing approximate solutions to the many-body problem. If previously approximate solutions were found, as a rule, by cutting off or partial summation of series, then the density functional method makes it possible to build approximations based on analogies with other, simpler systems, the solution for which is derived by independent methods. This highly physical approach is far from being exhausted and promises many interesting results in the future.

Concluding this lecture, we should note that modern models of the dynamics of nonideal plasma have reached today a sufficient level of perfection, making available a thermodynamic description of states with extremely high (concentration, solid) densities and ultrahigh pressures of the megabar–gigabar range.

References

[1] Al'tshuler L V 1965 Use of shock waves in high-pressure physics *Sov. Phys.-Usp.* **8** 52–91
[2] Al'tshuler L V, Trunin R F, Krupnikov K K and Panov N V 1996 Explosive laboratory devices for shock wave compression studies *Phys.-Usp.* **39** 539
[3] Alt'shuler L V, Brusnikin S E and Marchenko A S 1989 Determination of the Grüneisen coefficient of a strongly nonideal plasma *High Temp.* **27** 492–7
[4] Andersen O K 1975 Linear methods in band theory *Phys. Rev.* B **12** 3060–83
[5] Anisimov S I, Prokhorov A M and Fortov V E 1984 Application of high-power lasers to study matter at ultrahigh pressures *Sov. Phys.-Usp.* **27** 181–205
[6] Bakanova A A, Dudoladov I P and Sutulov Y N 1974 Shock compressibility of porous tungsten, molybdenum, copper, and aluminum in the low pressure domain *J. Appl. Mech. Tech. Phys.* **5** 241–5
[7] Car R and Parrinello M 1985 Unified approach for molecular dynamics and density-functional theory *Phys. Rev. Lett.* **55** 2471–4
[8] Dreizler R M and Gross E K U 1990 *Density Functional Theory: An Approach to the Quantum Many-Body Problem* (Berlin: Springer)
[9] Ebeling W, Fortov V E and Gryaznov V K *et al* 1990 Thermophysical properties of hot dense matter *Teubner-Texte zur Physik* (Leipzig: Teubner)
[10] Ebeling W, Kraeft W D and Kremp D 1976 *Theory of Bound States and Ionization Equilibrium in Plasmas and Solids* (Berlin: Akademie)
[11] Fortov V E 1982 Dynamic methods in plasma physics *Sov. Phys.-Usp.* **25** 781–809
[12] Fortov V E 2009 Extreme states of matter on earth and in space *Phys.-Usp.* **52** 615–47

[13] Fortov V E 2011 *Extreme States of Matter. Series: The Frontiers Collection* (Berlin: Springer)

[14] Fortov V E, Leont'ev A A, Dremin A N and Gryaznov V K 1976 Shock-wave production of a nonideal plasma *Sov. Phys. JETP* **44** 116–22

[15] Fortov V E, Lomakin B N and Krasnikov Y G 1971 Thermodynamic. Properties of a cesium plasma *High Temp.* **9** 789

[16] Fortov V E and Yakubov I T 1994 *Neideal'naya Plazma (Nonideal Plasma)* (Moscow: Energoatomizdat)

[17] Glukhodedov V D, Kirshanov S I, Lebedeva T S and Mochalov M A 1999 Properties of shock-compressed liquid krypton at pressures of up to 90 GPa *JETP* **89** 292–8

[18] Grigoryev F B, Kormer S B and Mikhailova O L *et al* 1985 Shock compression and brightness temperature of a shock wave front in argon. Electron screening of radiation *Sov. Phys. JETP* **61** 751

[19] Gryaznov V K, Ayukov S V and Baturin V A *et al* 2006 Solar plasma: calculation of thermodynamic functions and equation of state *J. Phys.* A **39** 4459

[20] Gryaznov V K, Fortov V E, Zhernokletov M V, Simakov G V, Trunin R F, Trusov L I and Iosilevski I L 1998 Shock compression and thermodynamics of highly nonideal metallic plasma *JETP* **87** 678–90

[21] Gryaznov V K, Iosilevski I L and Fortov V E 1996 Calculation of porous metal Hugoniots *Physics of Strongly Coupled Plasmas* ed W D Kraeft and M Schlanges (Singapore: World Scientific) pp 277–356

[22] Gryaznov V K, Iosilevskii I L and Fortov V E 1982 Thermodynamics of a highly compressed plasma in the megabar range *Sov. Tech. Phys. Lett.* **8** 592

[23] Gryaznov V K, Iosilevskiy I L and Fortov V E 2004 Thermodynamic properties of shock compressed plasmas based on a chemical picture *High-Pressure Shock Compression of Solids VII: Shock Waves and Extreme States of Matter* ed V E Fortov, L V Al'tshuler, R F Trunin and A I Funtikov (New York: Springer) pp 437–89

[24] Gryaznov V K, Iosilevskiy I L and Krasnikov Y G *et al* 1980 *Teplofizicheskie Svoistva Rabochikh Sred Gazofaznogo Yadernogo Reaktora (Thermophysical Properties of Working Media of a Gas-Phase Nuclear Reactor)* (Moscow: Atomizdat)

[25] Gryaznov V K, Iosilevsky I L and Fortov V E 1995 Raschet Termodinamicheskikh Svoystv Udarno-Szhatoy Plazmy Metallov (Calculation of the thermodynamic properties of a shock-compressed metal plasma) *Fizika Nizkotemperaturnoy Plazmy (Physics of Low-Temperature Plasmas)* (Russia: Petrozavodsk) p 105

[26] Gryaznov V K, Ivanova A N, Gutsev G L, Levin A A and Krestinin A V 1989 Programming package 'ESCAPAK' for calculations of electronic structure, adapted for the es series computers *J. Struct. Chem.* **30** 469

[27] Gryaznov V K, Zhernokletov M V and Zubarev V N *et al* 1980 Thermodynamic properties of a nonideal argon or xenon plasma *Sov. Phys. JETP.* **51** 288

[28] Hirschfelder J O, Curtiss C F and Bird R B 1954 *Molecular Theory of Gases and Liquids* (New York: Wiley)

[29] Hohenberg P and Kohn W 1964 Inhomogeneous electron gas *Phys. Rev.* **136** B864–71

[30] Iosilevski I L and Gryaznov V K 1981 On comparative accuracy of thermodynamic description of the properties of a gas plasma in the Thomas–Fermi and Saha approximations *High Temp.* **19** 799–803

[31] Iosilevskiy I L 1980 Equation of state of the non-ideal plasma *High Temp.* **18** 807

[32] Keeler R K, Van Thiel M and Alder B J 1965 Corresponding states at small interatomic distances *Physica* **31** 1437

[33] Kirzhnits D A, Lozovik Y E and Shpatakovskaya G V 1975 Statistical model of matter *Sov. Phys.-Usp.* **18** 649–72

[34] Kohn W 1999 Nobel lecture: electronic structure of matter - wave functions and density functionals *Rev. Mod. Phys.* **71** 1253–66

[35] Kohn W and Sham L J 1965 Self-consistent equations including exchange and correlation effects *Phys. Rev.* **140** A1133–8

[36] Kormer S B, Funtikov A I, Urlin V D and Kolesnikova A N 1962 Dynamical compression of porous metals and the equation of state with variable specific heat at high temperatures *Sov. Phys. JETP* **15** 477

[37] Kresse G and Joubert D 1999 From ultrasoft pseudopotentials to the projector augmented-wave method *Phys. Rev.* B **59** 1758–75

[38] March N H and Murray A 1961 Self-consistent perturbation treatment of impurities and imperfections in metals *Proc. R. Soc.* A **261** 119

[39] March N H and Tossi M P 1984 *Coulomb Liquids* (London: Academic)

[40] Marsh S P (ed) 1980 *LASL Shock Hugoniot Data* (Berkeley, CA: University of California Press)

[41] Martin R 2004 *Electronic Structure: Basic Theory and Practical Methods* (Cambridge: Cambridge University Press)

[42] Mintsev V B, Ternovoi V Y and Gryaznov V K *et al* 2000 Electrical conductivity of shock compressed xenon *Shock Compression of Condensed Matter-1999* ed S C Schmidt, D P Dandekar and J W Forbes (New York: Woolbury) pp 987–90

[43] Nellis W J, Van Thiel M and Mitchell A C 1982 Shock compression of liquid xenon to 130 GPa (1.3 Mbar) *Phys. Rev. Lett.* **48** 816–8

[44] Nikiforov A F, Novikov V G and Uvarov V B 2000 *Kvantovo-Statisticheskiye Modeli Vysokotemperaturnoy Plazmy i Metody Rascheta Rosseladnovykh Probegov i Uravneniy Sostoyaniya (Quantum-Statistical Models of High-Temperature Plasma and Methods for Calculating Rosseland Ranges and Equations of State)* (Moscow: Fizmatlit)

[45] Parr R G and Yang W 1989 *Density-Functional Theory of Atoms and Molecules* (New York: Oxford University Press)

[46] Perdew J P and Kurth S 1998 Density functionals: theory and applications *Lecture Notes in Physics* ed D Joubert (Berlin: Springer) p 8

[47] Pickard C J and Needs R J 2010 Aluminium at terapascal pressures *Nat. Mater.* **9** 624–7

[48] Radousky H B and Ross M 1988 Shock temperature measurements in high density fluid xenon *Phys. Lett.* A **129** 43

[49] Sin'ko G 1989 Opisanie Sistem Mnogikh Chastits Metodom Funktsionala Plotnosti (Description of many-particle systems by the density functional method) *Matematicheskoe Modelirovanie (Mathematical Modeling)* ed N N Kalitkin (Moscow: Nauka) pp 197–231

[50] Sin'ko G V 1983 Utilization of the self-consistent field method for the calculation of electron thermodynamic functions in simple substances *High Temp.* **21** 783–93

[51] Sin'ko G V and Smirnov N A 2009 *Ab initio* calculations for the elastic properties of magnesium under pressure *Phys. Rev.* B **80** 104113

[52] Sin'ko G V and Smirnov N A 2005 Relative stability and elastic properties of hep, bcc, and fcc beryllium under pressure *Phys. Rev.* B **71** 214108

[53] Slater J C 1937 Wave functions in a periodic potential *Phys. Rev.* **51** 846–51

[54] Slater J C and Koster G F 1954 Simplified LCAO method for the periodic potential problem *Phys. Rev.* **94** 1498–524

[55] Stefani F, Gundrum T and Gerbeth G *et al* 2006 Experimental evidence for magnetorotational instability in a Taylor-Couette flow under the influence of a helical magnetic field *Phys. Rev. Lett.* **97** 184502

[56] Trunin R F and Simakov G V 1993 Shock compression of ultralow density nickel *JETP* **76** 1090

[57] Trunin R F, Simakov G V and Sutulov Y N *et al* 1989 Compressibility of porous metals in shock waves *Sov. Phys. JETP.* **69** 580

[58] Urlin V D, Mochalov M A and Mikhailova O L 1992 Liquid xenon study under shock and quasi-isentropic compression *High Press. Res.* **8** 595–605

[59] Van Thiel M 1977 Compendium of shock wave data, Livermore Lawrence Laboratory *Rep. UCRL-50108*

[60] Vanderbilt D 1990 Soft Self-Consistent Pseudopotentials in a generalized eigenvalue fomalism *Phys. Rev.* B **41** 7892–95

[61] Vargaftik N B 1975 *Handbook of Thermophysical Properties of Gases and Liquids* (Washington: Hemisphere Publishing Corp.)

[62] Wang Y, Chen D and Zhang X 2000 Calculated equation of state of Al, Cu, Ta, Mo, and W to 1000 GPa *Phys. Rev. Lett.* **84** 3220–23

[63] Zel'dovich Y B and Landau L D 1943 On the relation of the liquid to gaseous state at metals *Acta Phys.-Chim. USSR* **18** 194

[64] Zel'dovich Y B 1957 Investigations of the equation of state by mechanical measurements *Sov. Phys. JETP.* **5** 1287

[65] Zhernokletov M V, Zubarev V N, Trunin R F and Fortov V E 1996 *Eksperimental'nye Dannye po Udarnoy Szhimayemosti i Adiabaticheskomu Rasshireniyu Kondensirovannykh Veshchestv pri Vysokikh Plotnostyakh Energii (Experimental Data on Shock Compressibility and Adiabatic Expansion of Condensed Matter at High Energy Densities)* (Chernogolovka: IPCP RAS)

[66] Zubarev V N, Podurets M A and Popov L V *et al* 1978 *Detonatsiya (Detonation)* (Chernogolovka: IPCP RAS) p 61

IOP Publishing

Lectures on the Physics of Extreme States of Matter

Vladimir E Fortov

Chapter 6

Lecture 6: Extreme states of matter in astrophysics

The physics of extreme states of matter underlies the modern understanding of the evolution and structure of the Universe [1].

According to modern ideas, the Universe consists of ordinary matter, photons, relic radiation, hidden mass, and 'vacuum-like' matter, which manifests itself as a nonzero cosmological constant [2, 3]. Ordinary matter is considered to mean mainly protons, electrons, and neutrons. Hydrogen is the dominant element. There is also helium and a small amount of lithium. Heavy atoms are found in very small amounts in the Universe. The number of protons in our Universe of radius 10^{28} cm is estimated to be $N = 10^{80}$, in agreement with the Eddington–Dirac number. The density of matter in the Universe is $\rho_{matter} = 10^{-31}$ g cm^{-3}. Ordinary matter is found in stars, planets, comets, interstellar gas, meteorites, and cosmic rays.

Anti-gravitating 'dark energy' makes up about 74% of all energy–mass [4]. The gravitating mass ('dark', or hidden, mass) accounts for about a quarter of the average density of the Universe, dark matter makes up 22%, and normal baryonic matter presented in the Mendeleev's Periodic table constitutes only about 4%. These 4% are found in stars, planets, and the interstellar medium. The interstellar medium accounts for 4/5 of the mass of baryonic matter and only 0.5% of the average density of the Universe is concentrated in stars. They occupy only a 10^{-25} part of the total volume of the Universe. Despite these modest 'average' figures, stars play a truly outstanding role in our Universe: their bright radiation reaches us from enormous distances and is the main source of information about various processes of matter–energy conversion in the Universe. In stars, there occur irreversible thermonuclear transformations, production of heavy elements, and generation of exotic, never-before-observed forms of matter, i.e. neutron matter, quark–gluon plasma, etc. While the information about the first hundreds of thousands of years of the evolution of the Universe reaches us in the form of relic radiation, we judge the history of the next billions of years by observing stars.

doi:10.1088/2053-2563/ab1091ch6

The range of changes in the parameters of matter in the Universe is extremely wide [5, 6]: from the cosmic vacuum and the rarefied intergalactic gas with a density of 10^{-30} g cm^{-3} (this value is accessed from the measurements of the gravitational effects of the vacuum and agrees with the upper limit which follows from the lower limit of the curvature of the space [7]) to extremely high densities of 10^{14}–10^{17} g cm^{-3} in neutron stars (table 6.1). The temperature of the intergalactic gas with a density $n \approx 10^{-4}$–10^{-3} cm^{-3} amounts to 10^7–10^8 K and may reach a billion degrees under heating by shock waves (from the shedding of the outer stellar layers, stellar collisions and explosions, collisions of gas clouds, etc). Inside the neutron stars, the temperature is 10^8–10^{11} K [1]. Over 99% of the visible matter is heated to a temperature exceeding 10^5 K.

While the magnetic fields are of the order of 10^{-9} and 10^{-6} G in the intergalactic space and near the Galactic plane, respectively, at the surface of neutron stars this field is 22 orders of magnitude higher. The record here belongs to recently discovered magnetars—neutron stars with a giant magnetic field of up to 10^{15} G, which corresponds to densities of the order of 10^8 g cm^{-3}, approaching the density of nuclear matter.

Giant black holes 'devour' entire star systems and hot galactic nuclei. The recently discussed magnetic tunnels ('wormholes') probably connect our and other universes, if they exist. Gravitational accretion of matter generates highly collimated jets, beams of charged particles accelerated to ultrahigh energies. Explosions of

Table 6.1. Characteristic parameters of matter in nature and in the laboratory.

Object	T, K	ρ, g cm^{-3}	p, bar
Interstellar gas	10^7–10^8	10^{-30}–10^{-3}	10^{-17}–10^{-7}
Earth, center	5×10^3	10–20	3.6×10^6
Jupiter, center	1.5–3×10^4	5–30	3–6×10^7
Exoplanets	10^3–10^5	1–30	10^7–10^8
Diamond anvils	4×10^3	5–20	5×10^6
Shock waves	10^7	13–50	5×10^9
Controlled thermonuclear fusion, magnetic confinement	10^8	3×10^{-9}	50
Controlled thermonuclear fusion, inertial confinement	10^8	150–200	2×10^{11}
Sun	1.5×10^7	150	10^{11}
Red giant	2–3×10^7	10^3–10^4	5×10^{12}
White dwarf	10^7	10^6–10^9	10^{16}–10^{22}
Relativistic collisions of gold nuclei, 100 GeV per nucleon, Brookhaven	2×10^7–7×10^{13}	10^{15}	10^{30}
Neutron star, black hole, γ-bursts	10^8–10^{11}	10^{14}–5×10^{15}	10^{25}–10^{27}
Early Universe (Planckian conditions)	10^{32}	10^{94}	10^{106}

supernovae generate shock waves, plasma emissions, turbulent plasma, and dust clouds, providing the raw material for the formation of new stars [5, 6, 8, 9].

The task of the researchers is to reproduce, to some extent, these exotic states and transformations of matter in the laboratory by colliding relativistic nuclei, in the focus of heavy-duty lasers, in colliding plasma pinches, or in supercomputer simulations [8]. At the same time, the difference in laboratory and astrophysical scales reaches many (up to 25) orders of magnitude; therefore, a choice of proper dimensionless variables and careful analysis of similarity criteria are needed [10].

The central element for describing the structure and evolution of astrophysical objects is the physical properties of a compressed and heated substance [1]. These are the equations of state of matter, composition of plasma, its optical properties, transport coefficients: viscosity, thermal conductivity, diffusion, electrical conductivity, bremsstrahlung of particles, and so on. This information is needed in a wide range of state parameters, only some of them being available today for laboratory measurements.

At the same time, theoretical models developed for a hot plasma or cold dense matter work well in fairly wide ranges [1, 8, 11, 12].

6.1 Planets, exoplanets, and low-mass stars

The detailed information received from automatic stations about the giant planets of the Solar System, as well as the discovery of hundreds of planets outside the Solar System (exoplanets), gave a significant impetus to planetary research [13]. The evolution and structure models developed here are based on quantitative information about the physical properties of compressed hot matter at megabar and ultramegabar pressures (figure 2.17, table 6.1).

6.1.1 Planets of the Solar System

Figure 6.1 shows the masses and sizes of the planets in our Solar System, and figure 6.2 [14] presents their sizes and the average distance from the Sun. The points in figure 6.1 are located almost on one line corresponding to an average density of about 3 g cm^{-3} [15], and lie in the range from 0.5 g cm^{-3} for comet nuclei to 7.7 g cm^{-3} for metal asteroids and meteorites.

The largest of the planets—Jupiter—is an order of magnitude smaller than the Sun, but has a density close to the Sun (1.33 and 1.41 g cm^{-3}, respectively). Saturn is close in size to Jupiter, but its density is almost 2 times less, i.e. 0.70 g cm^{-3}. The densities of Uranus and Neptune are 1.27 and 1.64 g cm^{-3}, respectively; together with Jupiter and Saturn, they form a group of giant planets of our Solar System. The Earth (with an average density of 5.52 g cm^{-3}), Venus (5.24 g cm^{-3}), Mars (3.94 g cm^{-3}), and Mercury (5.43 g cm^{-3}) are terrestrial planets with a high average density of matter. The circumsolar matter is 0.134% of the mass of the Solar System, the overwhelming amount of which (99.866%) is accounted for by the Sun, which is a typical yellow dwarf.

Figure 6.3 [14] presents the structure and characteristic parameters of some giant planets in the Solar System, in which 99.5% of the mass of the circumsolar matter is

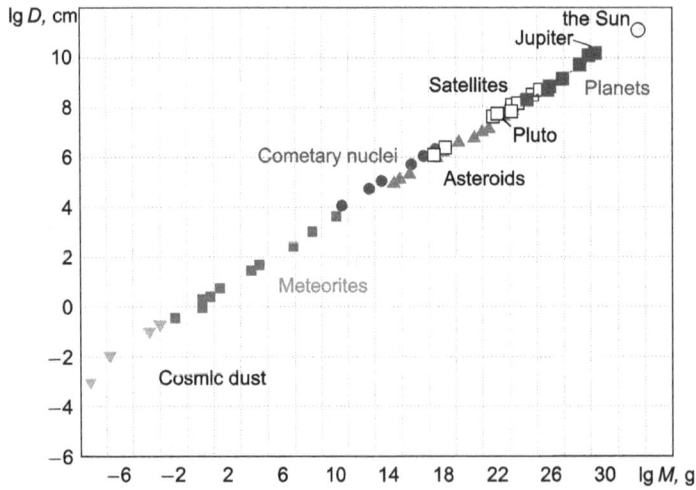

Figure 6.1. Masses and sizes of the objects of the Solar System.

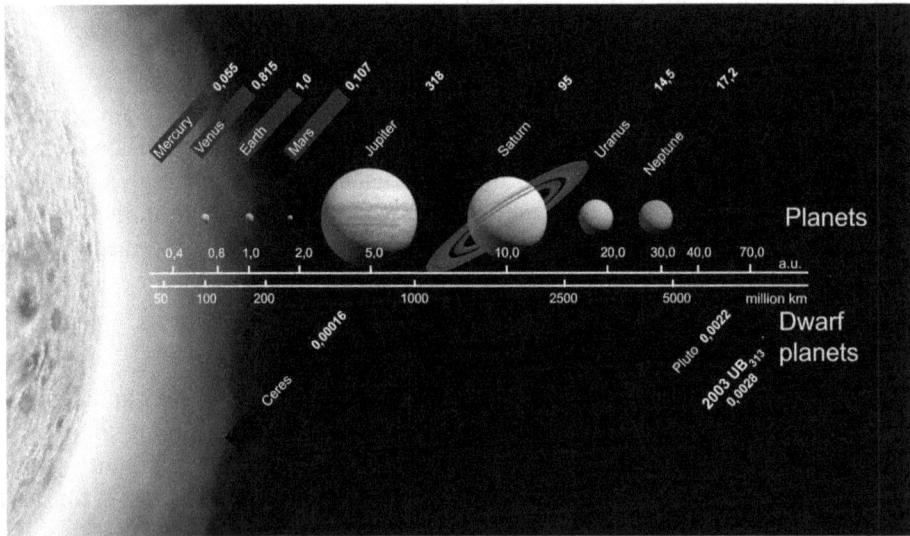

Figure 6.2. Masses of planets (in units of the Earth's mass) and their average distance from the Sun. Reprinted from [16] by permission from Springer. Copyright 2016.

concentrated. It is evident that we are dealing with a complex structure and various physical processes that we must figure out how to reproduce in the laboratory and describe by theoretical models of a dense plasma at megabar pressures.

Figure 6.4 shows the planetary matter phase diagram and the states available for the dynamic experiment, and figure 6.5 presents a diagram of the states of iron, where the states of the Earth and Hugoniot adiabats of iron are given. The internal structure of the Earth is illustrated in figure 6.6.

The solar mass is equal to approximately 2×10^{30} kg; the mass of the Sun is 333 000 times the mass of the Earth and 1000 times the mass of Jupiter.

Jupiter

Possible metallosilicate core

"Ices": water, ammonia, methane

Liquid molecular hydrogen + helium

At the center 70 Mbar, 20,000 K, 23 g/cm³

Liquid metallic hydrogen + helium

3 Mbar, 10,000 K

70,000 km

Saturn

"Ices": water, ammonia, methane

Liquid metallic hydrogen

Liquid molecular hydrogen

Gas-liquid atmosphere 0.69 Mbar 6,500 K

Metallosilicate core

Visible surface of the cloud layer

Gas-liquid atmosphere

At the center 45 Mbar, 10,000 K, 13 g/cm³

60,000 km

Uranus and Neptune

The Earth

Mantle

Lithosphere

Core

Mantle of "ices"

Metallosilicate core

At the center 4 Mbar, 6,000 K, 13 g/cm³

6,377 km

At the center 7-8 Mbar, 7,500 K, 10 g/cm3

~25,000 km

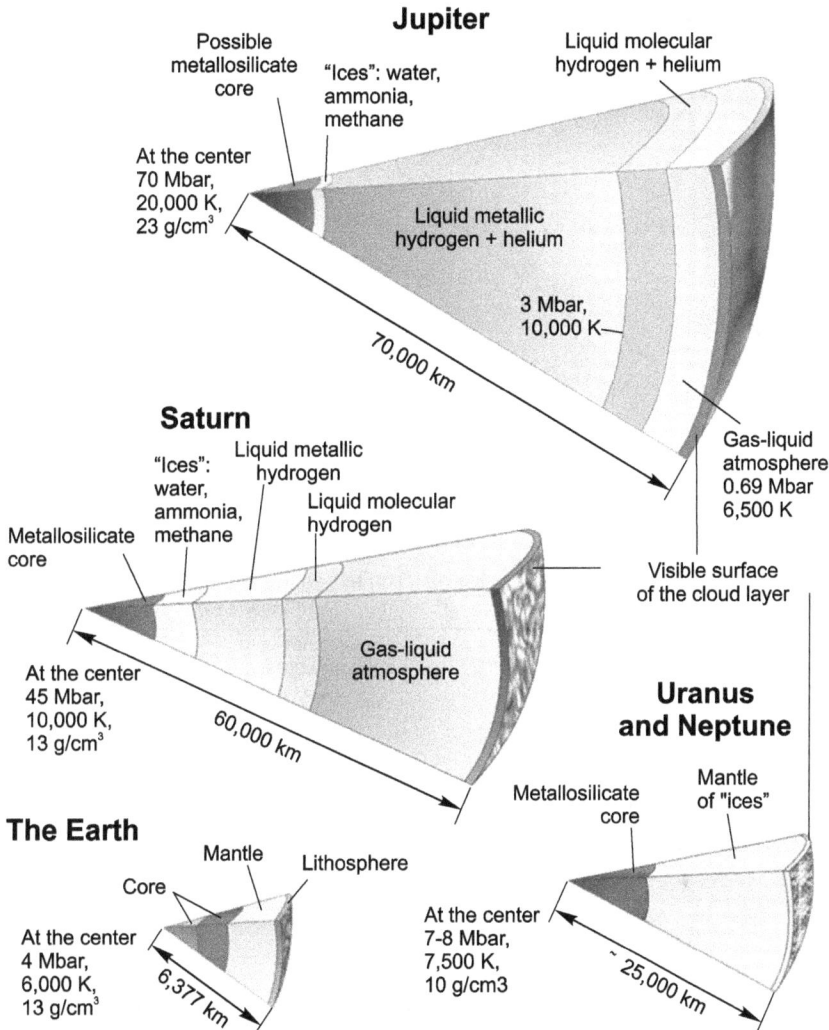

Figure 6.3. Scheme of the internal structure of giant planets in comparison with the structure of the Earth. Reprinted from [16] by permission from Springer. Copyright 2016.

The most massive stars known today are about 100 times the mass of the Sun. These massive stars are only 1500 times heavier than the most low-mass stars. In this case, star luminosities may differ more than a trillion-fold [4].

The sizes of stars differ not so much, but also significantly—almost by a billion times (if we do not take into account the neutron stars). In this case, the largest stars are not necessarily the most massive. There are stars that are larger than our Sun by about 1500 times in diameter. However, some of them do not differ significantly from the Sun in mass and have an average density millions of times smaller than the density of the Sun, despite the fact that the average density of the Sun is 1.4 g cm^{-3}, which is only slightly higher than the density of water [4].

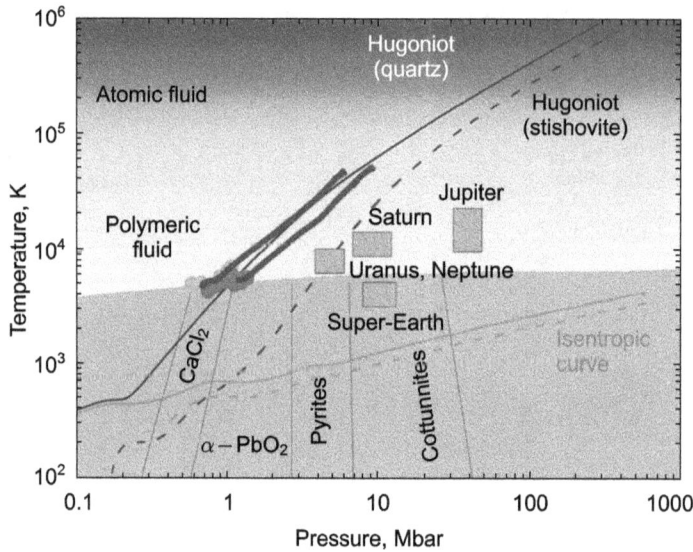

Figure 6.4. Phase diagram of matter. Reprinted from [16] by permission from Springer. Copyright 2016.

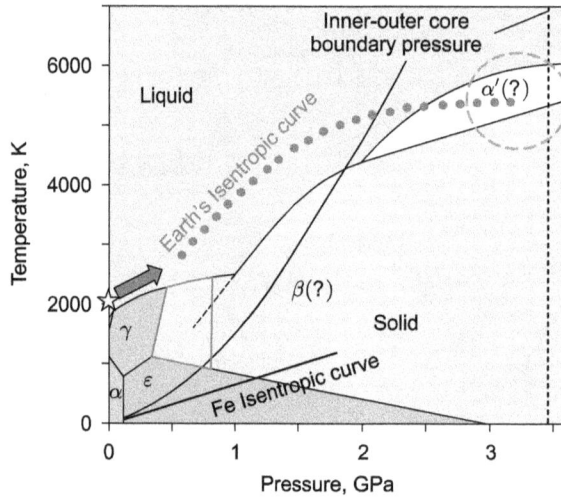

Figure 6.5. Phase diagram of iron with the Hugoniot adiabat and indication of terrestrial states. Reprinted from [16] by permission from Springer. Copyright 2016.

6.1.2 Exoplanets

We have far less observable information about the planets located outside our Solar System. Since 1992, several hundred such objects have been discovered, which are identified by astronomers from the lowering of the star brightness at the time the planet passes between it and the terrestrial observer [17]. A rare Hubble Space Telescope image of an exoplanet, a satellite of the star Gliese 229 [18], is shown in figure 6.7. The bright halo on the right is the optical exposure of the photodetector. This exoplanet has about 20–60 times Jupiter's mass.

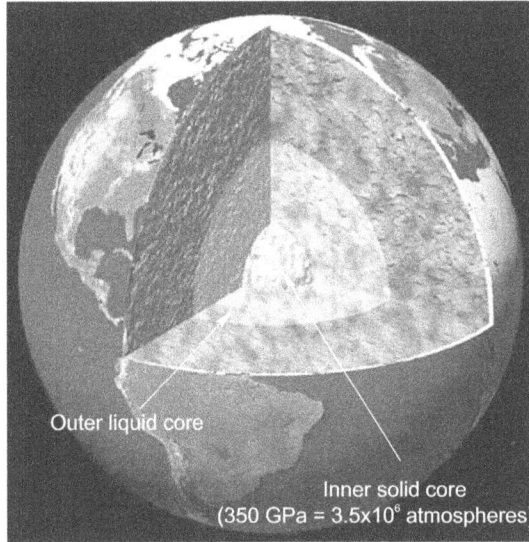

Figure 6.6. Internal structure of the Earth. Reprinted from [16] by permission from Springer. Copyright 2016.

Figure 6.7. Hubble Space Telescope image of Gliese 229 with its satellite. Reprinted from [16] by permission from Springer. Copyright 2016.

Only one of the discovered exoplanets, known as Gliese 581, is located at a distance of 20.5 light years from Earth and remotely resembles our planet. Its temperature is 15 °C–55 °C. The mission of NASA's Kepler spacecraft launched in 2009 is to find Earth-like exoplanets. The exoplanets discovered to date are very massive objects (figure 6.8 [13]) with a wide spectrum of orbit sizes, which determines various physical conditions of their matter.

For example, the exoplanet DH2094586 discovered in 1998 resembles Jupiter in its structure and parameters (somewhat exceeding its temperature) with a region of the plasma phase transition and the phase separation of hydrogen and helium plasmas. In most cases researchers predict the presence of a massive core (with a mass of up to 100 Earth masses) containing heavy elements. Figures 6.8 and 6.9 present the sizes and masses of some giant planets and exoplanets, as well as their chemical composition in comparison with several planets of the Solar System.

Figure 6.10 shows the abundance of the elements according to [13], where circles indicate compounds important for giant planets and exoplanets.

To interpret the data of terrestrial and space measurements, as well as to construct evolution, structure and energy models of these objects on their basis, reliable information about the physical properties of nonideal plasma in the megabar pressure range is required.

The case in point is a dense multicomponent plasma with strong collective interparticle interaction, where, along with the effects of thermal ionization, pressure-induced ionization, i.e. so-called 'cold' ionization, plays a determining role. Indeed, in the central region of Jupiter (see figures 6.3, 6.9, and 6.13), pressures reach 40–60 Mbar at $(15$–$20) \times 10^3$ K, while the plasma pressure at the center of the brown dwarf GL 229 is about 10^5 Mbar [9].

For these purposes, as we have noted, of particular interest are experiments on multiple (quasi-adiabatic) shock compression of hydrogen, helium, and their mixtures by pneumatic guns and explosive propellant devices [12, 19]. The experiments provide interesting and new information about astrophysical objects. These are data on the equation of state, on pressure-induced ionization (figure 6.11) [12] and on the phase transition (figure 6.12) [19] in nonideal plasma. Measurements have shown that the plasma phase transition is observed under isentropic compression of deuterium at $p \approx 1.2$ Mbar, where pressure-induced ionization occurs at a plasma density

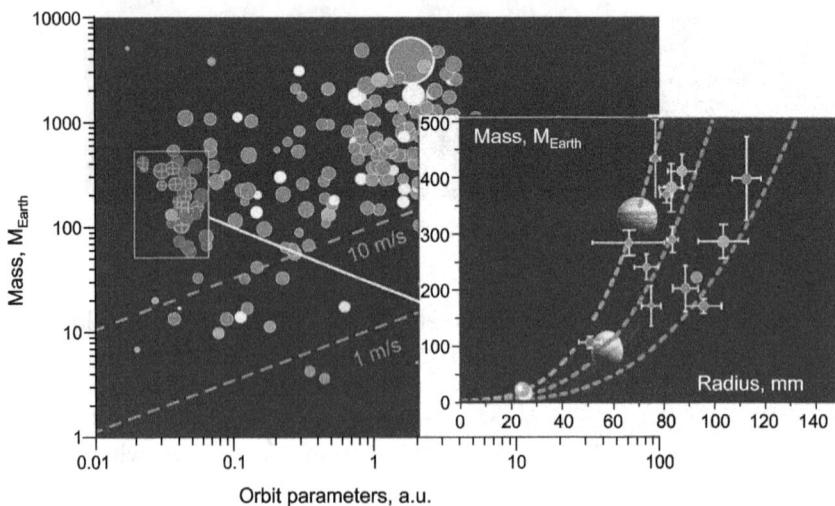

Figure 6.8. Characteristics of exoplanets. Reprinted from [16] by permission from Springer. Copyright 2016.

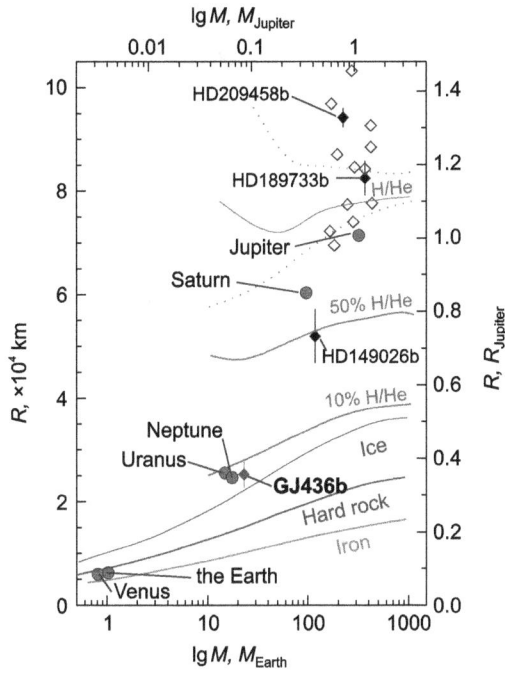

Figure 6.9. Characteristics and chemical composition of exoplanets in comparison with some planets of the Solar System. Reprinted from [16] by permission from Springer. Copyright 2016.

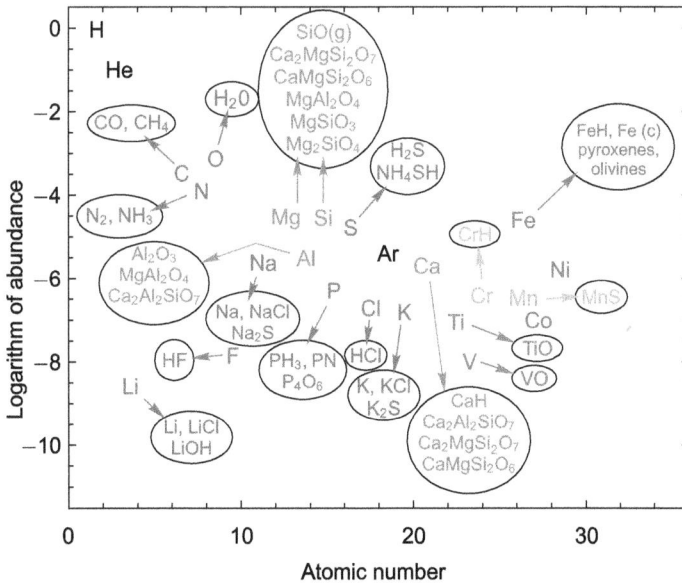

Figure 6.10. Abundance of chemical elements according to [13]. Reprinted from [16] by permission from Springer. Copyright 2016.

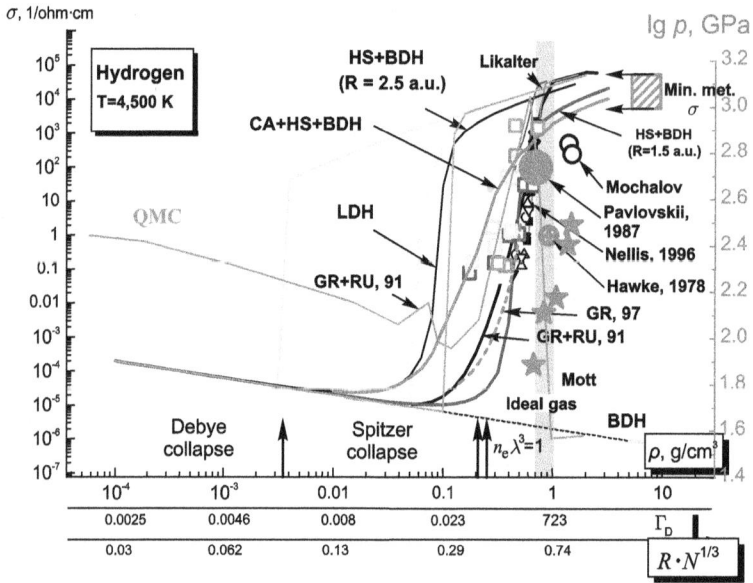

Figure 6.11. Pressure-induced ionization of a nonideal hydrogen plasma. The region of the thermodynamic phase transition is marked in gray [19], the asterisks indicate the results of density measurement by the pulsed x-ray radiography [19], QMC presents calculations by the quantum Monte Carlo method.

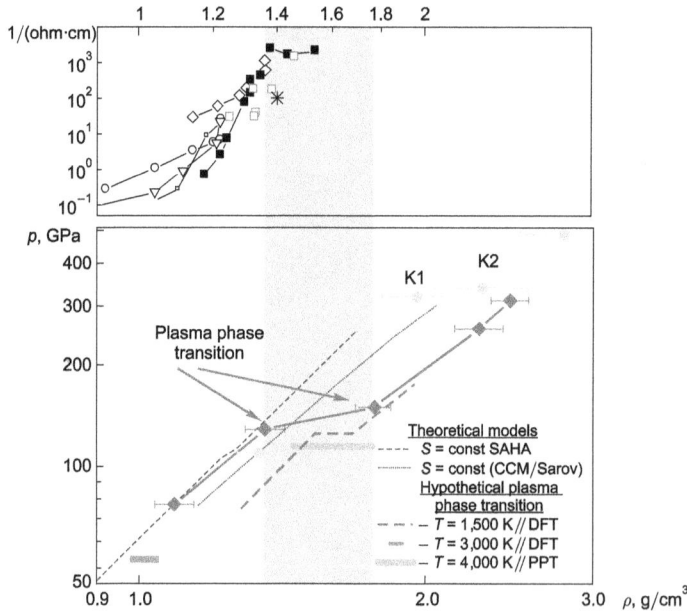

Figure 6.12. Adiabatic compressibility of a deuterium plasma [19]. The region of the plasma phase transition is marked in yellow. The data on the electrical conductivity in terms of the density of D2 is shown at the top (see figure 6.11). Reprinted from [16] by permission from Springer. Copyright 2016.

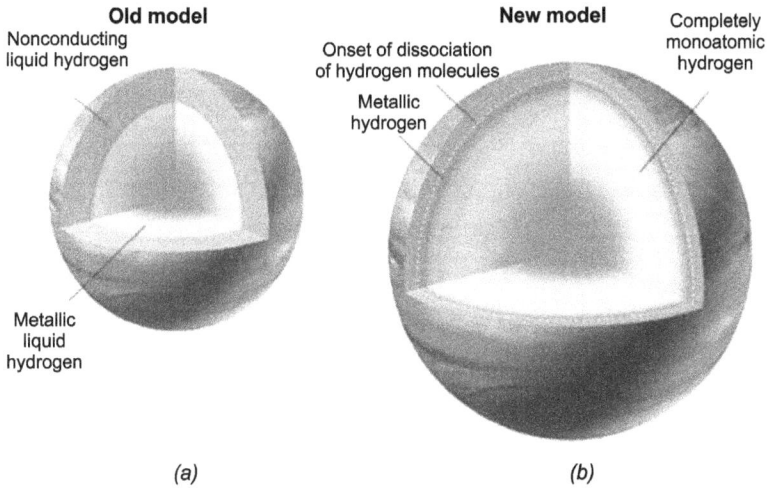

Figure 6.13. Scheme of the structure of Jupiter (a) before and (b) after measurements of electrical conductivity in shock-compressed hydrogen [20]. The zone of the metallic core has shifted from 0.75 to approximately 0.9 of Jupiter's radius (b). Reprinted from [16] by permission from Springer. Copyright 2016.

$\rho \approx 0.5\text{--}1.0$ g cm^{-3}. This permitted the radius in Jupiter at which metallization occurs to be measured, by shifting it toward large radii (figure 6.13) [11, 20].

The magnitude of the 'cold' ionization pressure is essential for the evaluation of convective phenomena and the generation of high (10–15 G) magnetic fields of Jupiter. The presence of a plasma phase transition is of interest for the assessment of internal (including gravitational) energy release in the phase separation of helium and heavy elements, as well as for the estimation of heat fluxes. Therefore, it is necessary [8] to further study the phase diagram of hydrogen and helium plasmas, hydrogen–helium mixtures, plasma phase transitions, metallization boundaries, mutual solubility of nonideal plasma of various composition and chemical elements, and the possibility of the appearance of 'helium rain' at high temperatures and pressures.

6.1.3 Low-mass stars and substars

Close in size to planetary objects are low-mass stars and substars, in the interior of which nuclear reactions have died out [21] due to insufficient mass. Usually their mass is 0.07–0.09 times the mass of the Sun and their size is comparable with that of Jupiter. These 'failed stars' range from Jupiter to the Sun in mass and consist of a hydrogen–helium degenerate or partially degenerate nonideal plasma with a pressure of about 0.1 Mbar in the center, while plasma in white dwarfs is completely degenerate.

As in the case of planets, the study and construction of atmospheric models of such substars is an extremely challenging task, requiring detailed thermodynamic and spectral calculations of molecular multicomponent plasma (millions of spectral lines and bands). It is also necessary to take into account the presence of condensates, the shift and broadening of the spectral lines, as well as the presence

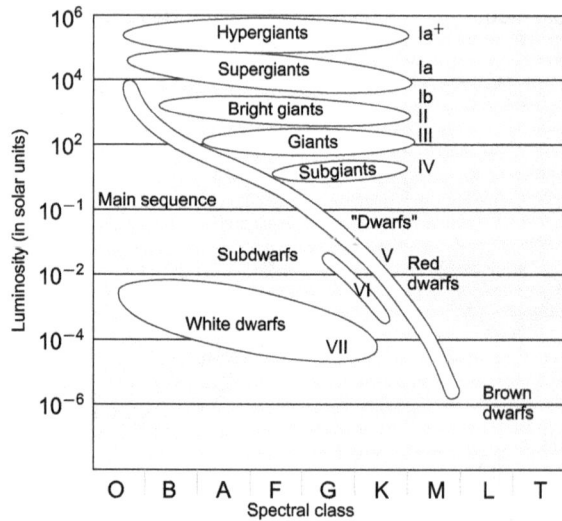

Figure 6.14. Schematic view of the Hertzsprung–Russell diagram in modern form. Regions occupied by main sequences of stars (groups of stars) are shown. To the right of the sequence name is its luminosity class, which is usually given together with the spectral class of the star.

of metals and their compounds. All this is needed, in particular, to calculate the luminosity of these poorly observed objects and to determine their contribution to the hidden mass of the Galaxy.

In general, the optical properties of the plasma of stars are one of the central areas of application of high-energy-density physics to astrophysics, since radiation determines the energy transfer inside the stars, their evolution, and the observed luminosity, providing the main observable information about these objects.

Radiative processes are of great importance for describing the origin, evolution, and structure of interstellar objects. The position of the stars on the luminosity–spectral class diagram is shown in figure 6.14 [4].

6.1.4 Brown and white dwarfs

The lower mass boundary of the stars is somewhat 'fuzzy', since small stars are difficult to observe because of their weak luminosity; for example, a red dwarf with a mass of $0.06 M_\odot$ has a temperature of only 2000 K. Examples of light stellar objects are the Ross 614 binary system with masses of the components equal to 0.059 and $0.051 M_\odot$ (solar masses) and the LHS 1047 system with one of the lowest mass components, which is estimated to be only $0.055 M_\odot$ [22]. Calculations show that in low-mass stars with a mass of less than $(0.07–0.1) M_\odot$, the temperature is low and insufficient for thermonuclear fusion, and the source of their energy is gravitational contraction, which is stopped by the pressure of a degenerate electron gas. The term 'star' is hardly applicable to low-mass objects without a thermonuclear energy source. The objects of mass equal to $(0.02–0.04) M_\odot$, adjoining this limit, are called brown dwarfs, which indicates that they emit infrared radiation. Using indirect methods, astronomers manage to observe lower-mass objects, such as stellar companions of mass $0.009 M_\odot$ in BD 68°946, which are objects intermediate between stars and planets [22].

The interior of brown dwarfs is composed of a fully ionized hydrogen–helium plasma. The degeneracy parameter $\psi = kT/kT_F$, where T_F is the Fermi temperature, ranges from 2 to 0.05 for very low-mass stars and brown dwarfs; therefore, the thermodynamic properties of a partially degenerate electron gas should be taken into account to describe the internal state.

Figure 6.15 illustrates the central characteristics of low-mass stars and substellar objects in the mass range from Jupiter to the Sun [23]. For low-mass star ($0.35 M_\odot$), the interior becomes completely convective, and since the gas still remains in the classical mode ($\psi \geqslant 1$), the mass increases in proportion to the star radius R, and the central density ρ_c decreases with increasing mass, as M^{-2}. Below the hydrogen burning limit, electronic degeneracy begins to dominate, and so $M \sim R^{-3}$, and the density again increases with increasing mass. The dependence of the radius on the mass $R \sim M^{-1/3}$, as in white dwarfs, takes place under complete degeneracy; however, when the degeneracy is partial, the dependence is more gently sloping because the radius of a brown dwarf depends slightly on the mass. A brown dwarf with the solar age and metallicity has a radius equal to that of Jupiter ($0.1 R_\odot$).

Figure 6.16 shows the temperature variation in the center of a low-mass object as a function of time.

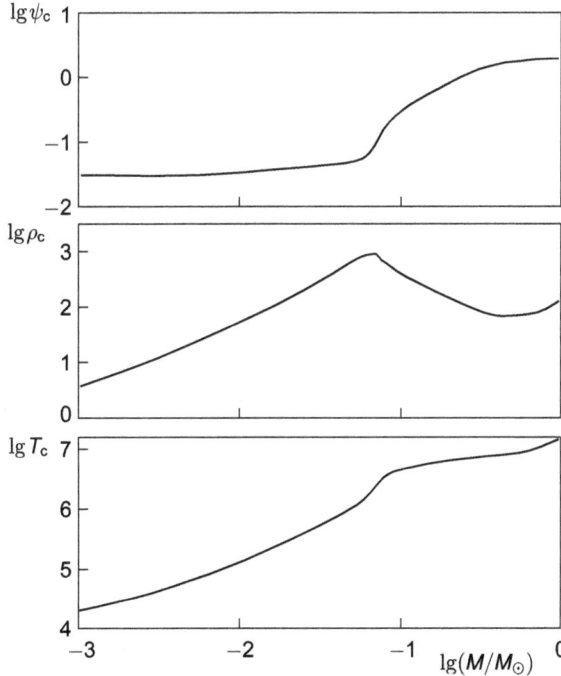

Figure 6.15. Degeneracy parameter ψ, central density ρ_c, and central temperature T_c for low-mass stars and brown dwarfs with a solar content of heavy elements and an age of 5 billion years.

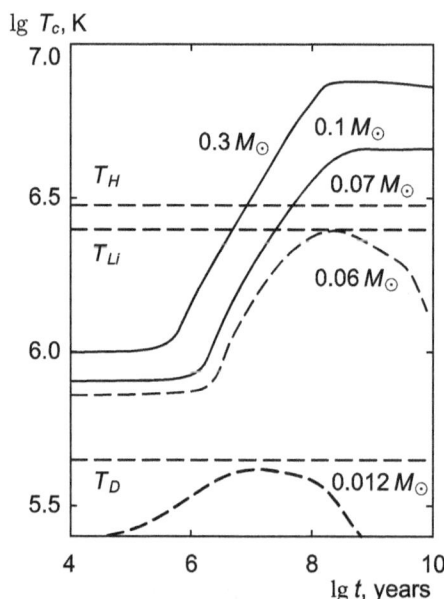

Figure 6.16. Central temperature T_c as a function of age for different masses. T_H, T_{Li}, and T_D are the temperatures of hydrogen, lithium, and deuterium burning.

The construction of atmospheric models for a brown dwarf is a challenging task, since it is necessary to calculate the molecular absorption (millions of spectral lines), to take into account the presence of condensates, etc.

Also, an important point is that the luminosity of brown dwarfs is 10^{-2}–10^{-5} below the solar luminosity. This greatly complicates the search for such low-luminous objects [23].

White dwarfs, the accretion of which may lead to type-Ia supernovae, are quite interesting astrophysical objects from the standpoint of realization of extreme states. In the evolution of stars of mass $(8-10)M_\odot$, the thermonuclear reaction stops at the stage of a helium or carbon–oxygen degenerate core. This thermonuclear burning in a degenerate core is explosive, and the increased temperature can partially remove the degeneracy and slow the energy release. Therefore, the outer shell of a red giant can be blown away because of the development of thermal instability at the shell source–degenerate core interface, thereby forming a planetary nebula. Interestingly, positive ions here form a crystal lattice [9], i.e. a kind of phase transition in nonideal plasma [19, 24].

A different kind of stellar activity arises if a white dwarf is part of a binary system where matter flows from a neighboring star to a white dwarf under the action of gravitational forces, thereby increasing the mass of the recipient. As the white dwarf's mass approaches the Chandrasekhar limit [9], thermonuclear fusion reactions are ignited at its center, leading to detonations (type Ia supernova explosion model). A very important problem arises in the development of explosion models: at the initial stage, a deflagration combustion wave is initiated. Eventually, it triggers carbon–oxygen detonation, followed by transition to iron-group elements with the predominance of the famous ^{56}Ni, which, decaying through ^{56}Co to ^{56}Fe, shapes the light curve of type Ia supernovae. However, in the case of oxygen–neon–magnesium (O–Ne–Mg) white dwarfs, this does not necessarily occur, since the neutronization of matter can begin prior to thermonuclear burning and then they may collapse into a neutron star.

The presence of white dwarfs in binary stellar systems manifests itself in a wide class of variable stars, which are referred to as explosive variables [9]. Their main distinguishing feature is the presence of periodic or irregular flares of varying amplitude, and the characteristic size of such systems ranges from fractions of to several solar radii.

Stars, in which explosions are relatively small and occur in the surface layers, are called nova-like stars. Known also are the stars with significant flare activity when explosion involves deeper layers of stellar interior (several percent of the radius). These stars are referred to as novae. Finally, when a big part of the star explodes, we are dealing with a so-called supernova.

The energy parameters of the explosions also differ markedly. For the 'weakest' observable astronomical explosions—flares on the Sun—the time scale t is about 10^3 s and the energy release E amounts to about 10^{32} erg. For novae, $t \approx 10^8$ s and $E \approx 10^{45}$ erg. For supernovae, $t \approx 10^{10}$ s and $E \approx 10^{50}$ erg. Probably, in galactic nuclei $t \approx 10^{15}$ s and $E \approx 10^{65}$ erg, etc.

The physical causes for the emergence of flares in various types of explosive variable stars are different. Single high-power flares, characteristic of novae, are caused by a thermonuclear explosion of matter accumulated on the surface of a white dwarf accreting from a neighboring main sequence star or from a solar-mass subgiant that underwent a minor evolution. Calculations show that single flares are possible only in a certain range of white dwarf masses and rates of matter accretion on their surface. At very low rates of flow, matter gradually degenerates and accretes on the white dwarf, while at very high rate, matter remains nondegenerate and combustion may proceed in a 'slack' quasi-stationary mode. For a nova to be stable, it is essential that degeneracy takes place in the material accreting on the surface, and the temperature increase is not accompanied by an increase in pressure and shell expansion, triggering a thermonuclear explosion. Since the gravitational energy of matter under these conditions is almost 100 times less than the energy of thermonuclear burning, particles in explosive thermonuclear burning acquire a velocity much greater than the escape velocity on the surface of the white dwarf, and the exploded shell disperses in the interstellar medium.

The kinetics of the evolution of white dwarfs requires comprehensive data on the equation of state of a plasma, its optical and transport properties, and is used for the temporal analysis of the corresponding galactic domain. Unfortunately, only outer regions of these objects can be presently observed experimentally.

At the same time, modern computer codes make it possible to carry out a meaningful numerical simulation of a supernova explosion. Figure 6.17 [10] presents the results of a two-dimensional simulation of a type II supernova explosion, where the development of hydrodynamic instabilities during the expansion of a plasma is clearly visible.

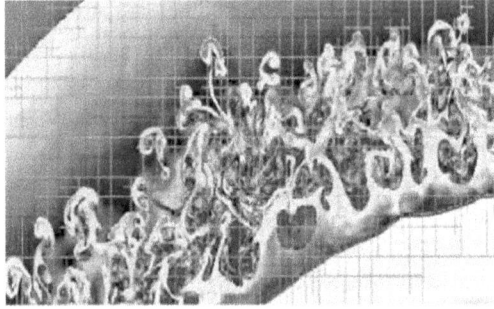

Figure 6.17. Two-dimensional numerical simulation of the SN1987A supernova explosion. Reprinted from [16] by permission from Springer. Copyright 2016.

In several scenarios for the evolution of stars (supernovae, novae), thermonuclear burning is initiated in the interior regions of a degenerate plasma and is then transferred to outer regions due to convection. In this case, convective instabilities develop, leading to an explosion of the object. Similar convective processes, but without a local thermonuclear energy release, are likely to occur in the outer regions of brown dwarfs.

Unfortunately, the convective effects in a degenerate plasma have now been studied insufficiently [8], which calls for appropriate laboratory experiments.

The existing high-pressure techniques—diamond anvils and intense shock waves driven by light-gas guns, electrodynamic installations, chemical and nuclear high explosives, lasers, and high-current pinches—allow one (see lecture 2) to study the equations of state, optical and transport properties of a plasma highly compressed under terrestrial conditions to record pressures of 4 Gbar [25–27], as well as to reach pressures of up to 10 Gbar inside laser microtargets so far without quantitative measurements of plasma properties. In some cases [19, 28], quasi-adiabatic compression is realized, sharply reducing the effects of heating the material.

It can be hoped that NIF and LMJ laser systems will make it possible to dramatically expand the available range of parameters and achieve the conditions typical of terrestrial planets, exoplanets, giant planets, brown dwarfs, and intermediate mass stars, as well as of the outer layers of white dwarfs.

6.2 Superextreme states of matter in compact astrophysical objects

Depending on the initial mass of the star with the solar chemical composition, three types of compact remnants may emerge in the stellar interior upon completion of thermonuclear evolution: white dwarfs, neutron stars, and black holes [1, 9].

As is known, massive stars die in a colossal explosion, after which a giant star, many times the size of the Sun, collapses down to a neutron star, i.e. a small (the size of a small asteroid) and extremely dense (denser than an atomic nucleus) fast-rotating ball with a strong magnetic field. The enormous force of gravity on the surface of a neutron star does not allow it to collapse, even when it rotates, like some radio pulsars, at a rate of about 1000 times per second (with a surface rotation speed of about 20% of the speed of light) [4].

6.2.1 Neutron and quark stars

Physicists consider a neutron star as one of the most interesting astronomical objects due to the set of its amazing properties. Despite the high temperature, its substance has superfluidity and superconductivity: the conditions necessary for these 'super-effects' arise because of the colossal density in the interior of a neutron star. From the point of view of plasma physics, extremely interesting is the atmosphere of such a star, where charged particles interact with an ultra-high magnetic field (up to 10^{14} G and even higher), which will never be obtained in terrestrial laboratories.

More than 2000 neutron stars have been discovered in our galaxy, and a few dozen more have been found outside of it. Basically, most of them are radio pulsars, and the rest are x-ray or gamma-ray sources. Astronomers believe that there are at least 100 million old neutron stars in the Galaxy that are difficult to find, because their surface has cooled and rotation has slowed down. The attracted gas spirals toward the neutron star at about 100 000 km s^{-1} and, striking the surface, is heated up to several million degrees. The resulting emissions of x-rays are detected by space-based observatories.

Neutron stars are the smallest observable stars in the Galaxy [29]. Their radii are about 10 km, which is 10^5 less than the size of ordinary stars and hundreds of times smaller than the radii of white dwarfs. However, the neutron star masses M are of the order of the solar mass $M_\odot$ and group at about $1.4M_\odot$. The average density of matter of neutron stars is $\bar{\rho} = 3M/(4\pi R^3) = 7 \times 10^{14}$ g cm^{-3}, i.e. exceeds the standard nuclear density $\rho_0 = 2.8 \times 10^{14}$ g cm^{-3} by several times. These are the densest of the celestial bodies known to us. The neutron star can be conditionally imagined as a huge atomic nucleus about 10 km in size. The density in the center of the star can reach values exceeding the nuclear density by a factor of 10–20. At such densities, condensation of pions, hyperons, and kaons is possible at the center of the neutron star. The possibility of the formation of strange quarks is also discussed.

The superfluidity of the baryonic matter component of neutron stars is one of the most interesting features of neutron stars. According to modern models, super-fluidity is due to Cooper pairing of baryons with opposite momenta under the action of an attractive component of strong interaction of particles. Superfluidity occurs at $T < T_c$ and gives rise to a gap Δ in the dispersion relation of baryons near the Fermi level.

The superfluidity of free neutrons, as well as of nucleons in atomic nuclei, was predicted in the inner core of the neutron star [30]. Neutrons, protons, and other particles may be superfluid in the core of a neutron star. Superfluidity of charged particles, for example protons, means superconductivity.

Hyperon matter allows hyperon pairing, whereas quark matter may give rise to quark pairing.

In these cases, the calculations yield critical temperatures $T_c \lesssim 10^{10}$ K and below. A new type of quark superfluidity is predicted, which consists in the pairing of different quarks (ud, us, ds). In this case, at a characteristic quark Fermi energy of about 500 MeV, one can expect critical temperatures $T_c \approx 50$ MeV $\approx 5 \times 10^{11}$ K.

Color-superconducting
strange quark matter

Surface
Hydrogen/He
Iron nuclei

Outer crust
Ions
Electron gas

Inner crust
Heavy ions
Relativistic electron gas
Superfluid neutrons

Outer core
Neutrons, protons
Electrons, muons
Superconducting protons

Inner core
Neutrons, protons
Electrons, muons
Hyperons (Σ, Λ, Ξ)
Boson condensate (π, K)
Deconfined quarks/quark matter

Radius ~10 to 14 km, Mass ~1 to 2$M_\odot$

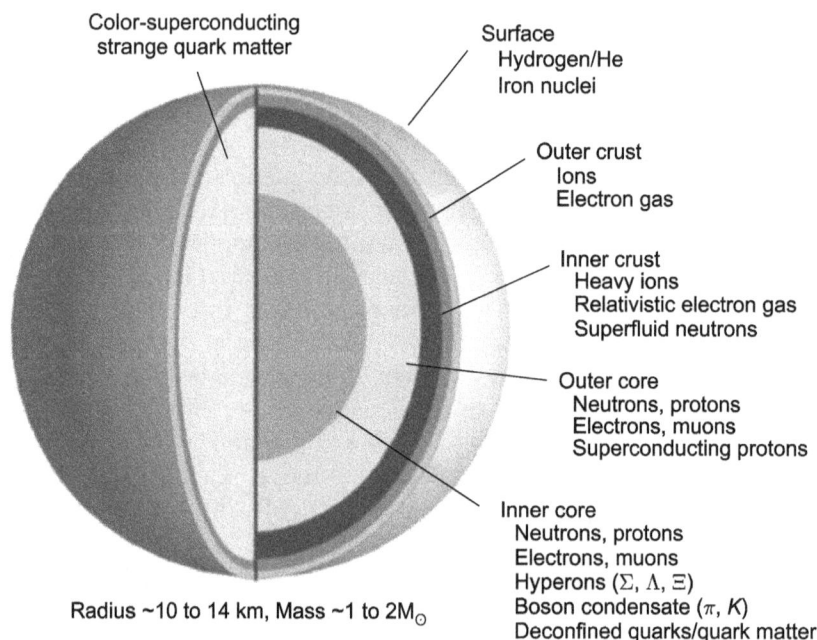

Figure 6.18. Structure of a hybrid or strange star based on the theory of nuclear matter. Reprinted from [16] by permission from Springer. Copyright 2016.

The further contraction of neutron stars may lead to deconfinement of quarks and the formation of a high-density quark–gluon plasma. The structure of such a hypothetical quark star is shown in figure 6.18 [31].

6.2.2 Magnetars

An important feature of relativistic astrophysical objects is the presence of giant magnetic fields, which largely determine the dynamics of their motion and radiative characteristics.

The radiation detected from a neutron star represents mainly soft x-rays and is reached at least 10^5–10^6 years after birth. Recently, a new class of neutron stars has been recorded, i.e. magnetars possessing a superstrong magnetic field up to 10^{15} G, which affects their gamma radiation formed together with the thermal radiation of the surface [9].

An example of such an object can be the brightest gamma-ray flare recorded in our galaxy to date; its source is the neutron star SGR 1806-20 belonging to the class of magnetars and located about 50 000 light years from the Earth. The flare lasted only about 0.1 s, and the amount of energy released within a few seconds thereafter was greater than the energy radiated by our Sun in 250 000 years.

The magnetic field of this magnetar amounts to 10^{15} G. SGR 1806-20 has turned to be the strongest of the known magnets in the Universe. For comparison, the value of the solar magnetic field varies from 1 to 5 G, and the Earth's magnetic field is only 0.31–0.62 G. If SGR 1806-20 were in place of the Moon, its magnetic field would

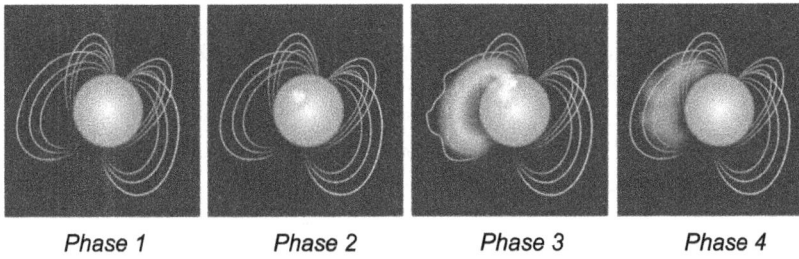

Phase 1   Phase 2   Phase 3   Phase 4

Figure 6.19. Magnetar flare origin. Reprinted from [16] by permission from Springer. Copyright 2016.

change the arrangement of molecules in all living organisms, not to mention radiation. But, fortunately, in the vicinity of the Sun there are no such extreme stellar objects (the nearest magnetar is 13 000 light years away). The magnetic field of SGR 1806-20 is several hundred times stronger than that of conventional radio pulsars. In this case, this star has a surface temperature of 10 million degrees Celsius. Due to this ultraintense magnetic field alone, a piece of iron at the surface of SGR 1806-20 should experience a force 150 million times stronger than Earth's gravity.

The extremely short duration of the flare indicates that the source of its energy is the dissipation of the magnetic field stored in the magnetosphere rather than in the core of the neutron star. The resulting expansion of the 'magnetic cloud' with a small amount of the baryon component (with $M < E/c^2$) is determined by the magnetic field. It is highly relativistic and anisotropic and retains these properties for several weeks after the flare.

Magnetars are known to show strong magnetic bursts; in this case, cracks are formed in the crust of the star under the action of the Lorentz force, and protons escape from these cracks, interact with the magnetic field, and emit energy. The magnetic field of a star is determined from the amount of this energy. Zasov and Surdin [32] proposed the following model of the magnetic flare of a magnetar: A magnetar is stable most of the time, but the stress induced by the magnetic field in its solid crust gradually increases (phase 1 in figure 6.19). At some point, the stresses in the crust exceed its ultimate strength, and it breaks into many small pieces (phase 2). This 'starquake' generates a pulsating electric current, which quickly decays, giving rise to a red-hot plasma sphere (phase 3).

The plasma sphere emits x-rays from its surface, cools down, and evaporates in a matter of seconds (phase 4).

The central element in understanding the structure and evolution of stellar objects, the collapse of stars, the explosion of supernovae, etc, is the equation of state, composition and optical properties of nonideal plasma of stellar matter [1, 12, 33, 34]. Their equation of state includes an important condition for nuclear statistical equilibrium, but it is applicable only for very high temperatures (about 5×10^{10} K). The corresponding equations of state are formulated for an arbitrary set of atomic nuclei (including unstable ones) and take into account the Coulomb effects of strong nonideality of matter, sometimes called the Coulomb liquid. Apart from the great complexities of describing a strongly nonideal plasma, there are additional challenges arising from the inclusion of a magnetic field, which hinders the correct description of the relation between the nuclear center-of-mass motion and the electronic structure. Only the first steps have been made in this direction. A phase transition of a nondegenerate gas into a state of macroscopic condensate has been predicted to occur at relatively low fields and temperatures [12, 19]. With increasing field and decreasing temperature, the density of saturated vapors of this condensate will become lower, lengthening optical paths and allowing thermal radiation to escape freely from the center of the star occupied by the condensate of a metallic liquid [8].

Today, a new generation of super-power short-pulse lasers already generates fields of about 10^9 G in a laser-produced plasma (see [35, 36] and lecture 3). Further advancement of this line of research can provide new information about emissivity and spectra of astrophysical plasmas in high-power magnetic fields.

An ingenious way of studying physical phenomena in superintense magnetic fields is discussed in [37]. For electron–hole plasma (liquid) in semiconductors, the magnetic fields, which are lower by a factor of $(\varepsilon m_0/m)^2 \approx 10^4$–$10^3$ (where m_0 is the proton mass and m is the hole mass) that for electron–proton plasma, will be superhigh. This makes electron–hole plasma a unique condensed object, which is available for laboratory investigations in superstrong magnetic fields.

6.2.3 Strange stars

As we saw in section 4.3, the quark–gluon plasma is an exotic superdense state of nuclear matter. This plasma has recently been discovered in the laboratory: it consists of light quarks, antiquarks, and gluons [38].

A quark–gluon plasma has a maximum density and can be generated in the center of neutron stars or in the collapse of ordinary stars (figure 6.20). In this case, we speak of quark and hybrid stars consisting of a quark–gluon core surrounded by a hadronic shell. In this case, the quark stars should be smaller in size than the neutron stars because of the greater compressibility of the quark–gluon plasma (figure 6.21). A candidate for a quark–gluon star or a hybrid star is, for example, the star RXJ1856 with a radius of more than 16 km, which was detected by the Newton and Chandra Space Telescopes in the x-ray range.

We note in this connection the hypothesis [39], according to which the plasma of almost free quarks is an absolutely stable state of matter not only at high pressures, but also at zero pressure. Should this be the case, a neutron star may turn into a so-called strange star, almost entirely consisting of quarks with a density of at least ρ_0.

Superdense matter states and, in particular, quark–gluon plasmas may also manifest themselves in black holes—objects predicted by the general theory of relativity, in which the gravitational field is so large that the escape velocity is equal to the speed of light [6, 40].

6.2.4 Black holes

The most 'extreme' states of matter, apparently comparable only with the very first stages of a warm Big Bang, are realized in black holes.

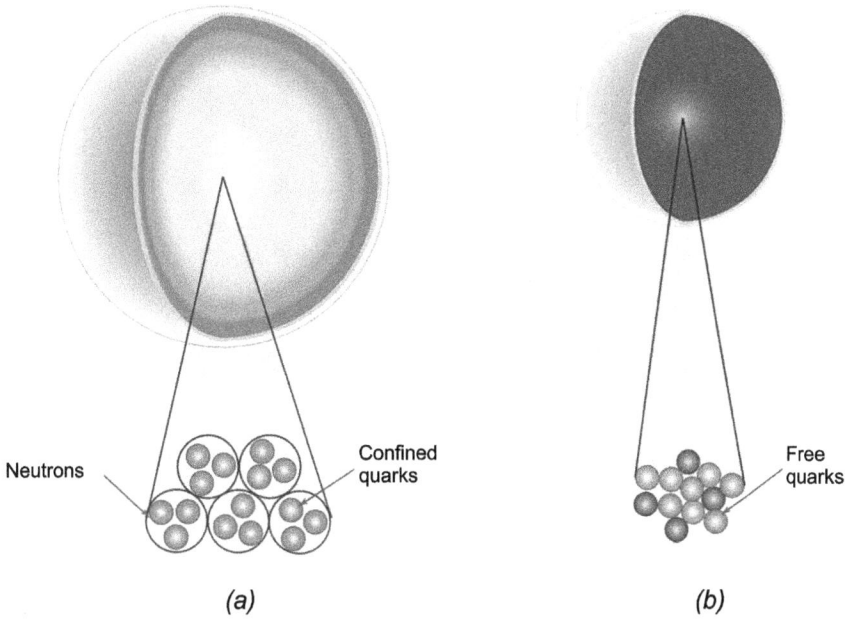

Figure 6.20. Neutron (a) and quark (b) stars. Reprinted from [16] by permission from Springer. Copyright 2016.

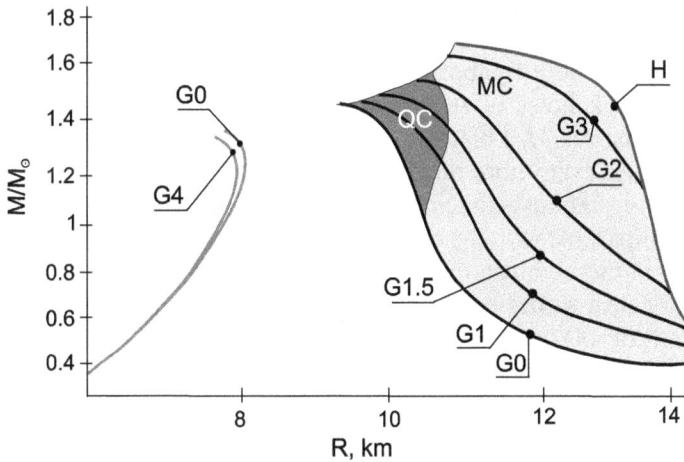

Figure 6.21. Sizes of quark and hybrid stars. Reprinted from [16] by permission from Springer. Copyright 2016.

The conditions for the formation of black holes correspond to the compression of matter to a certain critical density $\rho_{cr} = 2 \times 10^{16} \, (M_\odot/M)^2 \, \text{g cm}^{-3}$, which is inversely proportional to the mass of the object. For a typical black hole of stellar mass ($M = 10M_\odot$), the gravitational radius is 30 km, and the critical density amounts to $2 \times 10^{14} \, \text{g cm}^{-3}$ and is equal to the density of the atomic nucleus.

According to the existing ideas [3], our galaxy (observed by us as the Milky Way) can contain in its center a black hole with a mass of about $3 \times 10^6 M_\odot$ and a size of

$r/r_g = 1000$ $r/r_g = 10$

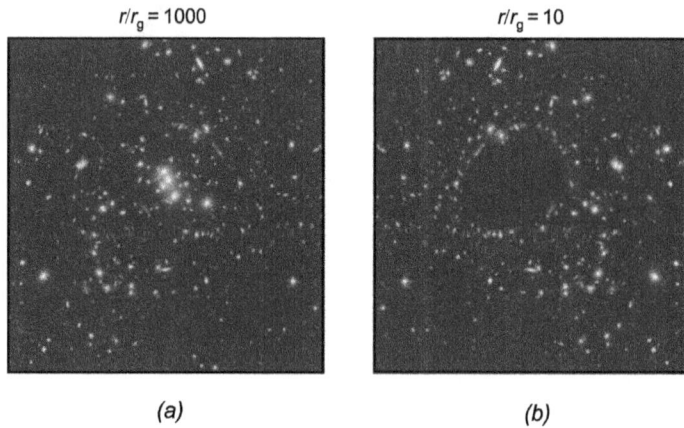

(a)   (b)

Figure 6.22. Black hole against the background of the starry sky. At a distance of $1000r_g$, its presence is hard to discern (a) and at a distance of $10r_g$, (b) it distorts the appearance of the starry sky. Reprinted from [16] by permission from Springer. Copyright 2016.

the order of 0.05 astronomical unit. The galaxy has a diameter of about 80 000 light years and a thickness of about 6000 light years; it also possesses a spherical halo with a radius of about 10^5 light years. The Sun and the Earth are 26 000 light-years away from the center of the Galaxy and orbit the Galaxy's center once every 230 million years, which makes only 22 rotations for 5 billion years of existence.

The presence of black holes can be judged indirectly from their gravitational action (figure 6.22) on neighboring objects. In our galaxy, there are about 20 candidates for black holes of stellar masses, while supermassive (of the order of billions of solar masses $M_\odot$) black holes are possibly present in about 300 galaxies. However, there are even more of these objects. Over the course of a long history, many stars, having exhausted their fuel, must collapse, so that the number of black holes can be comparable with the number of visible stars and, according to [41], their mass can explain the high rotational velocities of galaxies.

Recent years have witnessed the discovery of gigantic black holes with masses as great as millions of solar masses. Some of them existed even 13 billion years ago. The mechanism of formation of such giants is not yet clear. They either arose from the coalescence of black holes of smaller size or resulted from gravitational accretion.

High-current Z-pinches, lasers, and relativistic heavy ion accelerators (see lectures 2 and 3) can reproduce to a certain extent photoionization phenomena in plasma, allowing one to better understand the dynamics of the accretion of matter onto black holes.

The authors of papers [42, 43] came up with an interesting idea of producing mini-black holes under terrestrial conditions (figure 6.23). In this formulation, two exawatt lasers, irradiating gold targets, generate colliding ion fluxes with an energy of no less than 1 TeV, which should lead to the formation of a black hole with a radius of the order of 10^{-4} fm.

Figure 6.24 [44] shows the Hillas diagram with characteristic magnetic fields and sizes of astrophysical objects. It also demonstrates their capabilities as particle

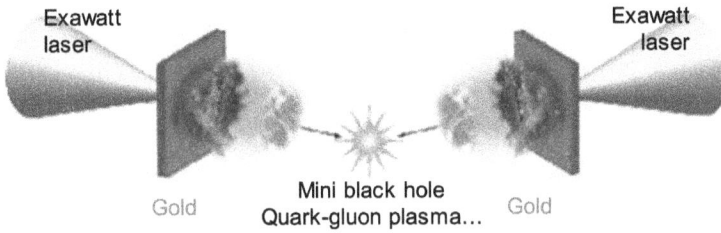

Figure 6.23. Laser generation of a quark–gluon plasma and a mini black hole. Reprinted from [16] by permission from Springer. Copyright 2016.

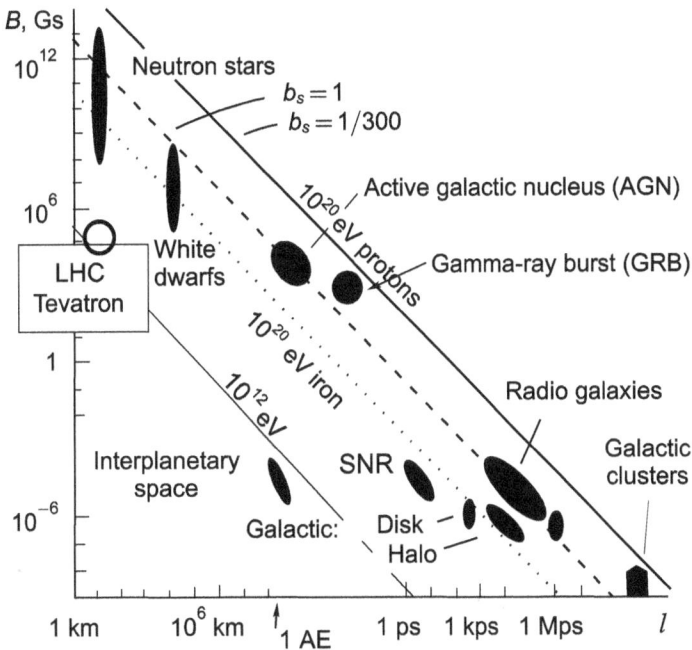

Figure 6.24. Hillas diagram with characteristic magnetic fields and sizes of astrophysical objects.

accelerators. Solid lines show the estimates of the sizes (in parsecs) and the values of magnetic field induction (in gausses) of objects capable of accelerating protons to energies of 10^{20} and 10^{21} eV at a shock wave velocity 300 times less than the speed of light. The dotted line represents the same for iron nuclei. Dark spots indicate the observed sizes and magnetic fields of various astrophysical objects.

It seems that in the Universe known to us, there are no, under the assumptions made about the particle acceleration mechanism (Fermi) itself, obvious candidates for the role of a Zevatron accelerator—an accelerator of observable-energy cosmic particles. The characteristics of the Tevatron and LHC accelerators are also shown for comparison.

References

[1] Arkani-Hamed N, Dimopoulos S and Dvali G 1999 Phenomenology, astrophysics, and cosmology of theories with submillimeter dimensions and TeV scale quantum gravity *Phys. Rev.* D **59** 086004

[2] Avrorin E N, Simonenko V A and Shibarshov L I 2006 Physics research during nuclear explosions *Phys.-Usp.* **49** 432

[3] Avrorin E N, Vodolaga B K, Simonenko V Λ and Fortov V E 1993 Intense shock waves and extreme states of matter *Phys.-Usp.* **36** 337–64

[4] Balega Y Y 2002 Brown dwarfs: substars without nuclear reactions *Phys.-Usp.* **45** 883–6

[5] Balega Y Y 2002 Brown dwarfs: substars without nuclear reactions *UFN* **172** 945–9

[6] Bleicher M 2007 How to create black holes on Earth *Eur. J. Phys.* **28** 509 16

[7] Burrows A, Hubbard W B, Lunine J I and Liebert J 2001 The theory of brown dwarfs and extrasolar giant planets *Rev. Mod. Phys.* **73** 719–65

[8] Cherepashchuk A M 1996 Masses of black holes in binary stellar systems *Phys. Usp.* **39** 759

[9] Cherepashchuk A M and Chernin A D 2004 *Vselennaya, Zhizn', Chernye Dyry (The Universe, Life, Black Holes)* (Fryazino: Vek 2)

[10] Chernin A D 2008 Dark energy and universal antigravitation *Phys.-Usp.* **51** 253

[11] Disdier L, Garconnet J P, Malka G and Miquel J-L 1999 Fast neutron emission from a high-energy ion beam produced by a high-intensity subpicosecond laser pulse *Phys. Rev. Lett.* **82** 1454–7

[12] Drake R P 2006 *High-Energy-Density Physics* (Berlin: Springer)

[13] Fortov V, Iakubov I and Khrapak A 2006 *Physics of Strongly Coupled Plasma* (Oxford: Oxford University Press)

[14] Fortov V E (ed) 2000 *Entsiklopediya Nizkotemperaturnoi Plazmy (Encyclopedia of Low-Temperature Plasma)* (Moscow: Nauka)

[15] Fortov V E 2007 Intense shock waves and extreme states of matter *Phys.-Usp.* **50** 333

[16] Fortov V E 2016 *High Energy Densities in Planets and Stars* (Berlin: Springer)

[17] Fortov V E, Ilkaev R I and Arinin V A et al 2007 Phase transition in a strongly nonideal deuterium plasma generated by quasi-isentropical compression at megabar pressures *Phys. Rev. Lett.* **99** 185001

[18] Friman B, Höhne C and Knoll J et al (ed) 2010 *The CBM Physics Book, Lecture Notes in Physics* vol 814 1 edn (Berlin: Springer)

[19] Ginzburg V L 1995 *O Fizike i Astrofizike (About Physics and Astrophysics)* (Moscow: Byuro Kvantum)

[20] Gribanov A A 2008 *Osnovnye Predstavleniya Sovremennoi Kosmologii (The Basic Representations of Modern Cosmology)* (Moscow: Fizmatlit)

[21] Haensel P, Potekhin A and Yakovlev D 2007 *Neutron Stars 1: equation of State and Structure* (New York: Springer)

[22] Hawking S W 1988 *A Brief History of Time: from the Big Bang to Black Holes* (Toronto: Bantam Books)

[23] Henderson D (ed) 2003 *Frontiers in High Energy Density Physics* (Washington: National Research Council, Nat. Acad. Press)

[24] Istomin Y N 2008 Electron–positron plasma generation in the magnetospheres of neutron stars *Phys.-Usp.* **51** 844

[25] Jeffries C D and Keldysh L V (ed) 1983 *Electron–Hole Droplets in Semiconductors* (Amsterdam: North-Holland)

[26] Kirzhnits D A 1972 Extremal states of matter (ultrahigh pressures and temperatures) *Sov. Phys.-Usp.* **14** 512–23

[27] Knudson M D, Hanson D L and Bailey J E *et al* 2001 Equation of state measurements in liquid deuterium to 70 GPa *Phys. Rev. Lett.* **87** 225501

[28] Mima K, Ohsuga T and Takabe H *et al* 1986 Wakeless triple-soliton accelerator *Phys. Rev. Lett.* **57** 1421–4

[29] Nellis W J 2002 Shock compression of hydrogen and other small molecules, in high pressure phenomena *Proc. of the Int. School of Physics 'Enrico Fermi' Course CXLVII* Chiarotti G L, Hemley R J, Bernasconi M and Ulivi L (Amsterdam: IOS Press) p 607

[30] Novikov I D 2001 'Big Bang' echo (cosmic microwave background observations) *Phys.-Usp.* **44** 817

[31] Panasyuk M I 2005 *Stranniki Vselennoi ili Ekho Bol'shogo Vzryva (Wanderers of the Universe or a Big Bang Echo)* (Fryazino: Vek 2)

[32] Rodionova Z F and Surdin V G 2007 Planety Solnechnoj Sistemy (Planets of the solar system) *Astronomiya: Vek 21 (Astronomy: 21st Century)* ed V G Surdin (Fryazino: Vek 2) p 34

[33] Samus' N N 2007 Peremennye Zvezdy (Variable stars) *Astronomiya: Vek 21 (Astronomy: 21st Century)* ed V G Surdin (Fryazino: Vek 2) p 162

[34] Shashkin A A 2005 Metal–insulator transitions and the effects of electron–electron interactions in two-dimensional electron systems *Phys.-Usp.* **48** 129

[35] Shevchenko V V 2007 Priroda planet (The nature of planets) *Astronomiya: Vek 21 (Astronomy: 21st Century)* ed V G Surdin (Fryazino: Vek 2) p 93

[36] Shevchenko V V 2000 Solnechnaya Sistema (The solar system) *Sovremennoe Estestvoznanie. Entsiklopediya (Modern Natural Science. Encyclopedia)* vol 4 ed V N Soifer (Moscow: Magistr Press) p 125

[37] Surdin V G (ed) 2009 Zvezdy (The Stars) 2nd edn *Astronomiya i Astrofizika (Astronomy and Astrophysics)* (Moscow: Fizmatlit)

[38] Surdin V G 1999 *Rozhdenie Zvezd (Star Production)* (Moscow: Editorial URSS)

[39] Trunin R F 1994 Shock compressibility of condensed materials in strong shock waves generated by underground nuclear explosions *Phys.-Usp.* **37** 1123

[40] Vacca J R 2004 *The World's 20 Greatest Unsolved Problems* (Upper Saddle River, NJ: Prentice Hall)

[41] Witten E 1984 Cosmic separation of phases *Phys. Rev.* D **30** 272–85

[42] Yakovlev D G 2001 Superfluidity in neutron stars *Phys.-Usp.* **44** 823–6

[43] Zasov A V and Postnov K A 2006 *Obshchaya Astrofizika (General Astrophysics)* (Fryazino: Vek-2)

[44] Zasov A V and Surdin V G 2007 Raznoobrazie Galaktik (A variety of galaxies) *Astronomiya: Vek 21 (Astronomy: 21st Century)* ed V G Surdin (Fryazino: Vek 2) p 329

IOP Publishing

Lectures on the Physics of Extreme States of Matter

Vladimir E Fortov

Chapter 7

Conclusion

The science of the structure of matter at extremely high pressures and temperatures and cosmophysics are closely related and interwoven (see Kirzhnits D A 2006 *Lektsii po Fizike* (*Lectures on Physics*) (Moscow: Nauka)). On the one hand, the solution of almost any cosmophysical problem is impossible without invoking the information about the structure of matter of the corresponding celestial object. On the other hand, cosmophysics provides both nuclear and subnuclear physics with information that significantly complements the data obtained in terrestrial laboratories. This applies, for example, to characteristics of nuclear forces (pulsar data) and the number of neutrino species (cosmological data).

A significant increase in the role of the Cosmos (especially the Universe as a whole), as a source of fundamental information, should be expected in the future as well. This is so because the limits of the potential of accelerator physics are already in sight. At the same time, the constantly emerging capabilities in the experimental physics of extreme states of matter give hope for the laboratory reproduction of ultra-extreme states of matter, so typical for the surrounding Universe.

Although the limiting pressures of a laboratory plasma are still 20–30 orders of magnitude higher than the maximum astrophysical values, this gap is being rapidly bridged, and the physical processes in a laboratory and in space often demonstrate an astonishing variety and at the same time striking similarities, evidencing at least the uniformity of physical principles of the behavior of matter in an extremely broad range of densities (approximately 42 orders of magnitude) and temperatures (up to 10^{13} K).

This all defies the most vivid imagination, and, as pointed out by Voltaire, '…nothing in nature is simpler or more orderly. The sovereign states of Germany or Italy, which one can traverse in a half hour, compared to the empires of Turkey, Muscovia, or China, are only feeble reflections of the prodigious differences that nature has placed in all beings' (Voltaire 2015 *Romans—Volume 3: Micromegas* (London: The Orion Publishing Group Ltd)).

www.ingramcontent.com/pod-product-compliance
Lightning Source LLC
Chambersburg PA
CBHW080545220326
41599CB00032B/6363